NATIONAL GEOGRAPHIC KiDS

ULTIMATE ROCK-OPEDIA

THE MOST COMPLETE ROCKS & MINERALS REFERENCE EVER

STEVE TOMECEK

NATIONAL GEOGRAPHIC

WASHINGTON, D.C.

TABLE OF CONTENTS

CHAPTER 7:
The Building Blocks
of Rocks 134

CHAPTER 8:
Rockin' Resources 248

INTRODUCTION

My love affair with rocks began in summer 1961 in a place called Pottsville, Pennsylvania.

I was five years old and visiting my uncle Joe's farm at the time. There were a lot of coal mines around Pottsville, and I remember hiking up in the hills looking at those shiny black rocks that people collected and burned. My uncle explained that the coal that they were digging up came from plants that grew in big swamps that had covered the area millions of years ago, even before the dinosaurs were alive. I thought he was making it all up. After all, the whole area was nothing but rocky mountains, and there were no swamps anywhere in sight. Then, one day, I kicked a big flat rock and it split open on the ground in front of me. And that's when it happened: Inside, there it was—the perfect print of a fern leaf ... right there in the rock! Uncle Joe hadn't been lying. I had discovered my first fossil, and I was hooked!

After that, every time I went for a walk in the woods or along a stream, I was more interested in looking at the rocks on the ground than in watching where I was going. (Yes, I got really wet and walked into a lot of trees!) I started collecting rocks everywhere I went and, when I got home, I put them in boxes with little labels. I went to the library and found lots of books about rocks. And, whenever possible, I would go to the American Museum of Natural History in New York City with my mom. Other kids went to look at the dinosaurs or the giant blue whale (which, to be honest, were pretty cool, too), but most of the time I went straight to the hall of rocks and minerals.

I loved learning about how the different rocks formed and what they could tell you about the history of an area. By the time I got to middle school, I knew that I wanted to be some type of geologist. When I finally graduated college and started working as a soil specialist and hydrogeologist, I was in heaven. Being a geologist was a lot like being a detective, but instead of hunting for clues about who committed a crime, I got to hunt for clues about how the Earth around me had changed and how to keep people from damaging the environment.

After many years of working with rocks, minerals, fossils, and (my personal favorite) soil, I decided that I wanted to share my love of geology with other people who might not realize how important rocks really are. That's why I wanted to write this book! It's actually the fifth book that I've written for the folks at National Geographic, but this one digs deeper than the others: *Ultimate Rockopedia* has tons of facts and figures about the rocky world around us, including profiles of rad rocks, marvelous minerals, and fantastic formations in all corners of the globe. You'll get the inside story on the features of our dynamic Earth, including volcanoes, earthquakes, and hydrothermal vents. You'll journey to the top of the tallest mountains and to the deepest place on Earth. Hopefully you'll enjoy reading the book and will come away with a love of rocks, too.

Keep rockin'!

Steve "The Dirtmeister" Tomecek,
geologist and author of
Ultimate Rockopedia

When I was a kid growing up in Brooklyn, New York, a trip to the American Museum of Natural History opened up a whole new world to me.

It was filled with dinosaur fossils that had been around for millions of years and hundreds of different rocks and minerals from all over the world—even some from outer space! There were so many amazing specimens of all colors, shapes, and sizes, I couldn't believe it. I became hooked on the subject that brought them all together: geology.

Weekend trips to the Catskill Mountains, exploring sandy beaches and rocky jetties, all reinforced my fascination with rocks and Earth's processes. In college, my fellow geology major rock hounds and I drove around collecting fossils and minerals from all over, even stopping along brand-new roads to hunt for unique and freshly dug-up finds.

When I became a teacher, there was nothing more awesome than hearing the *oohs* and *aahs* of my students as I cracked open a drab-looking geode to reveal the sparkling, colorful crystals that were hidden inside or as I split apart a rock to reveal a fossil for the first time. Sharing my passion with my students for more than 30 years was a hands-on experience, as I took them to the same places I had fallen in love with.

I spent my whole life chasing down specimens in the wild and studying geology. What's great about *Ultimate Rockopedia* is that it brings all of that fun, excitement, and wonder together for future generations of rock hounds. It will get you hooked, just like that kid in Brooklyn got hooked by seeing his first fossil in the American Museum of Natural History. Back then I would have loved to have had a reference book just like this to help fuel my passion.

This book is a gem, and I bet you'll really dig it!

Jim Lucarelli,
expert consultant for
Ultimate Rockopedia

MY NIECE AND ME AT THE AMERICAN MUSEUM OF NATURAL HISTORY

ON THE ROCKS

There are about 1,500 potentially active volcanoes on Earth.

Alaska has more earthquakes than any other U.S. state.

Earth's tectonic plates are moving continuously at about the same rate that human fingernails grow.

The largest known space rock on the surface of Earth is in Namibia. Weighing an estimated 66 tons (60 t), the iron-nickel Hoba meteorite was discovered in 1920 by a farmer when he was plowing his field.

The presidential heads carved in the granite at Mount Rushmore in South Dakota, U.S.A., are eroding at a rate of about one inch (2.5 cm) every 10,000 years.

Yellowstone National Park, in the western United States, sits on top of a massive supervolcano.

Royalty in ancient China were buried in suits made of jade because it was believed the mineral would help preserve and protect their bodies in the afterlife.

A fulgurite is a tube-shaped rock structure that forms when lightning strikes sand, causing the grains to melt and fuse together.

CHAPTER **ONE**

THERE'S NO PLACE LIKE **HOME!**

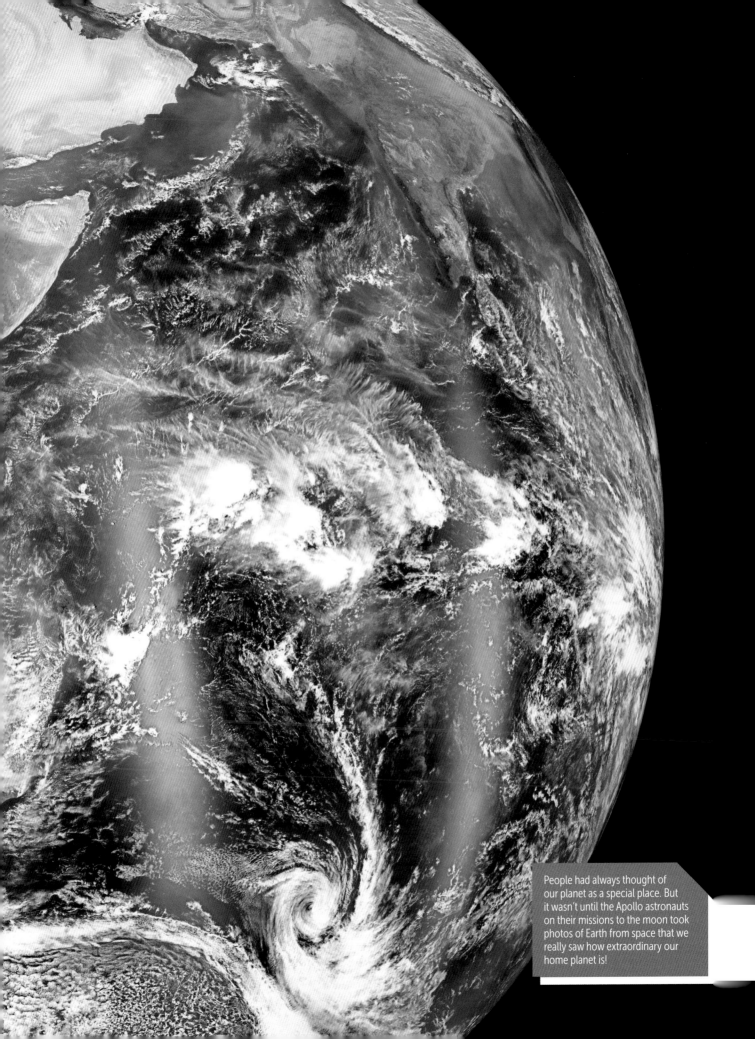

People had always thought of our planet as a special place. But it wasn't until the Apollo astronauts on their missions to the moon took photos of Earth from space that we really saw how extraordinary our home planet is!

OUR ROCKY WORLD

From space, you can clearly see that Earth is different from other planets in our solar system. The blue water of the ocean and the white clouds floating in the atmosphere form a stark contrast with the rocky green-and-brown continents. Early on, people believed that those rocks stopped at the edge of the sea, but we know now that rocks aren't found only where you can see them. Rocks can be found in every environment on Earth, from the tops of the highest mountains to the deepest depths of the ocean. They are buried beneath the soil and trees of the rainforests and lie under the frozen glaciers in Antarctica and Greenland. Rocks even provide the materials used to construct the foundations of the pavement and buildings in the world's greatest cities.

Over the years, scientists have made lots of important discoveries about the rocks that make up our world. They've discovered that there are different types of rocks that can form in different ways. They've also discovered that some rocks are really old and date back to a time when Earth was quite young, while others are forming at this very second. Most monumentally, scientists have learned that rocks preserve a record of the changes that have happened to our planet over time. By learning how to "read the rocks," geologists have been able to piece together the remarkable history of our planet.

Scientists aren't the only people interested in rocks, though. People all over the world depend on rocks for an incredible range of resources. Without rocks, there would be no metals or materials to make things like cell phones and computers. Rocks also provide some of the fuel to keep factories and power plants running, and they're used in the construction of buildings and roads. Most important, rocks together with decaying plants allow soil to form, and without soil, there would be no plants or animals or people on our planet.

When you stop and think about it, our world really does rock! Join us as we unlock some of the greatest mysteries on (and in!) Earth.

SOME OF THE **METEOROIDS, ASTEROIDS,** AND **COMETS** THAT OCCASIONALLY CRASH INTO OUR PLANET ARE LEFTOVERS FROM THE **ORIGINAL NEBULA** THAT OUR **SOLAR SYSTEM** FORMED FROM.

EROS ASTEROID

Bet You Didn't Know!

One of the earliest rocks that formed on Earth may have been found on the moon! Scientists believe that a four-billion-year-old rock fragment—brought home by Apollo 14 astronauts in 1971 along with 93 pounds (42 kg) of moon rocks and soil—was blasted off our planet and landed on the lunar surface when a huge asteroid hit Earth some time in the distant past. After studying the rock, scientists realized the minerals that it is made from are a better match for Earth rocks than for those that formed on the moon. This discovery was announced in 2019.

EL CAPITAN, A VERTICAL
ROCK FORMATION IN
YOSEMITE NATIONAL PARK,
IN CALIFORNIA, U.S.A.

HOW IT ALL
BEGAN

Most scientists believe our Earth, along
with the rest of our solar system, formed
about 4.54 billion years ago from a large
spinning cloud of gas and dust called
a nebula. Under the force of gravity, the
tiny particles in the nebula started joining
together to make clumps, which collided
with other clumps to make even bigger
clumps. The nebula eventually flattened
out to become a large spinning disk, with
the largest clump ultimately becoming
the sun sitting at the center of the solar
system. The rest of the clumps eventually
settled into stable orbits around the sun
and became the planets—including Earth,
the big clump that we call home!

THE ROSETTE
NEBULA IN THE
CONSTELLATION
MONOCEROS

WHAT IS A ROCK?

Most people know rocks when they see them, but when asked to explain exactly what a rock is, many people have a difficult time coming up with an answer. You might begin by saying a rock is hard and solid, but there are lots of things in the world that are hard solids that aren't rocks. A piece of wood, a brick, a chunk of concrete, and a coconut are all hard solids, and none of them are rocks.

There are a few things that make a rock stand out from the rest of the matter found on Earth. Most rocks are made of minerals. Minerals are kind of like the building blocks of rocks. Some rocks, like granite (pp. 78–79), are made from lots of different minerals joined together, while rocks like quartzite (pp. 102–103) and amphibolite (pp. 104–105) usually have only one type of mineral in them.

So, rocks are made from minerals. But what exactly is a mineral? A mineral is a naturally occurring inorganic (nonliving) solid that has a regular internal arrangement of atoms and molecules. The reason that a piece of wood and a coconut can't be rocks is because they are organic and were once alive. But what about a chunk of concrete? It's not alive. And it's actually made from little pieces of rock, so why isn't it a rock? Well, because materials like brick and concrete are made by humans, they are not naturally occurring, so even though they may contain minerals, they are technically not rocks.

So far, geologists have identified about 5,000 different kinds of minerals on our planet, each with its own unique combination of chemical elements. Some minerals, such as painite, are quite rare and are found in only a few locations in the world. Others, like quartz (pp. 206–207) and feldspar (pp. 212–213), are found all over the planet. The truth is that even though thousands of different minerals have been identified, only about 100 or so are commonly found making up most of the rocks around us.

One of the most common minerals found on Earth is ice. That's right, the stuff that you skate on in winter is a solid and has a regular internal structure of molecules with a definite chemical composition. Since it is also naturally occurring and nonliving, ice has all the required properties to make it one cool mineral!

GRANITE

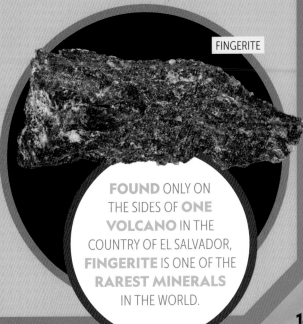

COAL

RULE BREAKERS OF THE ROCK WORLD

The world is filled with rule breakers. Take the platypus, for example. It lays eggs, but it is still classified as a mammal! There are a few rocks that break the rules, too—rocks that are not made from minerals but are still considered rocks. One example is coal, which formed from dead plants that got buried in the ground millions of years ago and slowly turned into a solid mass. Another example is coquina, a rock made entirely from broken pieces of seashells that became cemented together. Since both are made from things that were once alive, they technically shouldn't be called rocks—but because both occur naturally and are found in the ground, they are.

COQUINA

FINGERITE

AMPHIBOLITE

FOUND ONLY ON THE SIDES OF **ONE VOLCANO** IN THE COUNTRY OF EL SALVADOR, **FINGERITE** IS ONE OF THE **RAREST MINERALS** IN THE WORLD.

STACKING
STONES

Long before people built with concrete, steel, and glass, they used rocks, which were easy to come by, super strong, and very durable. Check out these iconic stone structures that have really stood the test of time.

THE MOAI OF EASTER ISLAND

Located in the Pacific Ocean, Easter Island—also known as Rapa Nui—is famous for the approximately 900 giant stone heads that cover the hillsides. Known as moai, these massive statues, which are up to 65 feet (20 m) tall and can weigh in at a whopping 14 tons (12.7 t), were carved by the early Rapa Nui people using stone picks made from basalt (p. 90). The main bodies were made from individual blocks of a rock called tuff (p. 91), and many have a red stone "hat" on top of their heads made from a volcanic rock called scoria (p. 91).

PYRAMID OF THE SUN

Sometime around A.D. 200, the Teotihuacan people of what is now the country of Mexico completed one of the largest structures ever built in the ancient world. Located about 30 miles (50 km) northeast of Mexico City, and standing more than 200 feet (61 m) high, the four-level Pyramid of the Sun dominates the local landscape. It is made of several different types of stones, including a local red volcanic rock called tezontle. Archaeologists believe that there was once a temple at the top of the structure that could be reached by climbing the 248 stone steps that run up one side.

GREAT PYRAMID OF KHUFU

The Great Pyramid in Giza, Egypt, was built more than 4,500 years ago as a tomb for the pharaoh Khufu. It is the largest pyramid ever constructed, originally standing 481 feet (147 m) high with a base that measures 755 feet (230 m) on each side. The pyramid itself is made from more than two million individual limestone (pp. 120–121) and granite (pp. 78–79) blocks, some weighing more than two tons (1.8 t) each. Recent discoveries suggest that it took a group of about 20,000 workers more than 20 years to construct!

STONEHENGE

The great stone circle known as Stonehenge has stood on England's Salisbury Plain for more than 5,000 years, from when the outer circle, made of giant sandstone (p. 127) blocks, was first built. These outer stones stand about 18 feet (5.5 m) high and weigh around 25 tons (23 t) each. But these are small compared with the horseshoe-shaped ring of stone structures in the center, the largest of which weighs more than 45 tons (41 t). That's as much as three fully loaded school buses!

MACHU PICCHU

At nearly 7,900 feet (2,400 m) above sea level, Machu Picchu is one of the best-preserved cities from the Inca Empire. Built about 500 years ago in the Andes Mountains of Peru, in South America, the site features more than 200 separate buildings, miles of walkways, and terraces built from precisely cut blocks of granite (pp. 78–79). Archaeologists are still working out the history of Machu Picchu, but they do know that it was used as an astronomical observatory to track the motions of the sun and moon.

MUCH **ADO** ABOUT
MINERALS

With thousands of different minerals making up the planet, how do geologists keep track of them all? Fortunately, minerals have several important properties that allow us to figure out which mineral is which.

AMETHYST

IT'S ALL ABOUT
BONDING!

Even though they are not alive, minerals are born and they grow! Minerals first form when individual atoms that are free to move around begin to link up in a process called chemical bonding. Bonding is not a random process, though. There are certain rules that control which types of atoms can join with one another and how they can fit together. As more and more atoms come together, the internal structure of a mineral begins to take shape. It's this internal arrangement of atoms and molecules that not only controls the shape of the mineral's crystals but also the way it breaks, the way it shines, how hard it is (based on the Mohs scale, p. 137), and even its color.

Bet You Didn't Know!

If you look at common table salt (which is the mineral halite) with a magnifying glass, you'll see that the individual grains usually have square edges. That's because when the sodium and chlorine atoms bond together to make the mineral halite (pp. 186–187), they form a cubic crystal structure.

PYRITE

CRYSTAL HABIT—Different minerals have different shapes. Some minerals, like the garnets (pp. 242–243), usually look a little like soccer balls, while others, like pyrite (pp. 160–161), normally form cubes. The shape that a mineral's crystal naturally forms is called its crystal habit. Minerals don't always form perfect crystals, though. Sometimes, their crystals can be misshapen—or even nonexistent! But when minerals do form crystals, their crystal habit is one of the best clues for figuring out what they are.

LUSTER—Some minerals really shine, while others look kind of dull. A mineral's luster describes the way that light reflects from, or bounces off of, a mineral's surface. A perfect diamond (pp. 152–153) really sparkles and is said to have a brilliant luster, while pyrite (pp. 160–161) is said to have a metallic luster because it shines like polished metal. Often, the terms used to describe a mineral's luster come from other common materials and include things like glassy, earthy, pearly, silky, and waxy.

COLOR AND STREAK—You might think that a mineral's color would be a great property to help identify what it is, but this is not always the case. Some minerals, like sulfur (pp. 140–141), have really distinctive colors. Others, like pyrite (pp. 160–161), always come in the same color. In these cases, the color of the mineral can be very helpful. But with a mineral like quartz (pp. 206–207), which can come in lots of different colors, it is not helpful at all. Another problem is that there are lots of minerals that have the exact same color so it can be easy to confuse them for one another. Instead of using color, geologists use a property called streak to help tell minerals apart. A mineral's streak is the color that the mineral leaves behind when it is rubbed against a special white unglazed porcelain tile called a streak plate. Unlike a mineral's outside color, the streak a mineral produces is much more distinctive, and it is always the same. That makes streak a much better property than color when trying to figure out what a mineral is.

THE MINERAL **GYPSUM** OFTEN FORMS **CRYSTALS** THAT LOOK LIKE **LITTLE FLOWERS** THAT GEOLOGISTS CALL **ROSETTES.**

GYPSUM

19

SOME REALLY CLASSY CRYSTALS

Archaeologists aren't sure exactly when people first started collecting gemstones, but the practice goes back well over 10,000 years. At first, people probably just picked up rocks with unusual colors or special shine, just like you might do when you are walking down a gravel path or along a lakeshore. Later, people learned which types of rocks some of these gems came from and went out of their way to start digging them up.

So what sets gemstones apart from ordinary rocks? First, they feature a dazzling luster or deep, rich colors. Second, most gemstones stand the test of time because they are usually very hard and durable. Finally, most gem-quality stones tend to be rare, although this is not always the case. Both garnet (pp. 242–243) and amethyst (pp. 208–209) are pretty common minerals, but they are still considered to be gemstones because of their color and the way they shine when cut. Gems are usually classified as being either precious or semiprecious depending on how common they are. Because they are relatively rare, diamonds (pp. 152–153), emeralds (pp. 230–231), rubies (pp. 168–169), and sapphires (pp. 168–169) are all considered to be "precious" gems, while the other gems are "semiprecious."

THE GREAT RUBY WATCH

One of the most important things to understand about gemstones is that they don't usually start out looking the way they do when they are in a piece of jewelry or decorating a golden goblet. Most gemstones get their "bling" only after they have been cut and polished. Polishing removes all the rough spots on the gem so that light can perfectly reflect off its surface. And cutting produces different angles, causing the gem to sparkle. The person who does the work of cutting and polishing gems is called a lapidary. Lapidaries have to be incredibly skilled at what they do because one mistake when cutting a diamond can cause it to shatter like a piece of glass!

THE HOPE DIAMOND

GEM-MAKING MINERALS

People are often surprised to discover that many gemstones are just fancy forms of some rather common minerals. Rubies and sapphires, for example, are just colorful forms of the superhard mineral corundum (pp. 166–167). And aquamarine and emerald are really brightly colored forms of the mineral beryl (pp. 230–231). Quartz (pp. 206–207), one of the most common minerals on Earth, gives us the gemstones amethyst and citrine, and the gem peridot is really just a rare form of the mineral olivine (pp. 234–235). All in all, there are about 20 common minerals that can produce gemstones. So, you may be wondering, what mineral is responsible for diamonds? Diamond is its own mineral—so you only get diamonds from another diamond!

THE HOOKER EMERALD

THE IMPERIAL STATE CROWN, ONE OF THE CROWN JEWELS OF THE UNITED KINGDOM

AFTER IT WAS DISCOVERED, THE **CULLINAN DIAMOND** WAS CUT INTO MORE THAN **100 SEPARATE DIAMONDS,** SOME OF WHICH ARE IN THE **CROWN JEWELS** OF THE UNITED KINGDOM.

Bet You Didn't **Know!**

The largest white diamond discovered to date was the Cullinan diamond, which originally weighed in at an incredible 3,106 carats. So what does this mean in average everyday units? One carat is equal to 0.2 gram, which is less than a small paper clip. That means the Cullinan diamond (before it was cut) weighed about as much as an average coconut.

ANCIENT
ARTWORK

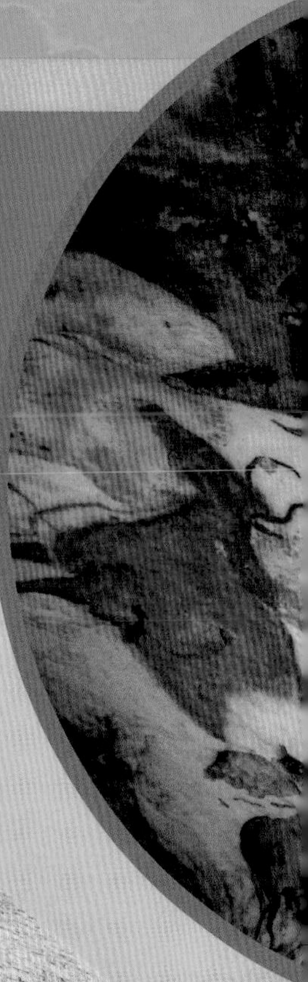

When it came to expressing their artistic abilities, people of the past used rocks in a bunch of creative ways. Here are a few ancient works of art that totally rock and are still amazing people today!

THE SPHINX

Standing guard in the desert sands of Giza, Egypt, the Great Sphinx is one of the largest stone statues ever created. The Sphinx, carved from a single block of limestone (pp. 120–121), has the body of a lion and the face of a person. Many archaeologists believe this particular face belongs to the pharaoh Khafre, who lived more than 4,500 years ago. Measuring 240 feet (73 m) long and 66 feet (20 m) high, the Sphinx is an example of rock art on a colossal scale!

THE NASCA LINES

Out across a dry plain near the city of Nasca, Peru, are a series of strange lines cut into the desert floor. The oldest lines were made by people more than 2,000 years ago and had baffled scientists until they observed them while flying above the area in an airplane. It turns out that what looked like random lines on the ground are actually gigantic drawings of animals and plants, some of which extend for hundreds of feet across the ground.

LASCAUX CAVE PAINTINGS

In 1940, four boys were exploring a large hole in the ground in southern France when they stumbled upon a curious cave. The walls of the ancient subterranean shelter—used by people more than 15,000 years ago—were covered with hundreds of color paintings and engravings of animals such as birds, bison, deer, and horses. It is one of the earliest art galleries ever discovered.

THE SUN DAGGER PETROGLYPHS OF CHACO CANYON

In 1977, an artist named Anna Sofaer was exploring the Chaco Canyon area of New Mexico, U.S.A., when she made a startling discovery. Behind three slabs of sandstone (p. 127) were two spiral patterns cut into the rock. She determined that these strange carvings were made by the ancestral Puebloan people—who had lived in the area more than 1,000 years ago—to help them keep track of the changing seasons. The slabs of sandstone created a single dagger of sunlight in the center of the large spiral on the summer solstice and two daggers on either side of the spiral during the winter solstice.

THE ROSETTA STONE

Dating back to around 200 B.C., this broken slab of granodiorite is one of the most important archaeological discoveries of all time. The stone has the same text inscribed in three different forms of writing—ancient Greek, hieroglyphics (a type of picture writing used in ancient Egypt), and cursive hieroglyphics. Using their knowledge of ancient Greek, historians were able to decipher the hieroglyphics. This made it possible for them to translate the writings on other ancient Egyptian artifacts—allowing them to learn a great deal about ancient Egyptian civilization.

CHAPTER TWO

EARTH IS
ROCKIN'

At first, scientists believed that our planet was a stable, unchanging world. Over the past 200 years or so, however, geologists have gotten a better understanding of just how dynamic Earth is. In the process, they've discovered that forces both inside and outside the planet continuously change its surface.

DYNAMIC PLANET

Over the past few decades, astronomers have learned a great deal about the other members of our neighborhood in space, otherwise known as the solar system. Uncrewed probes have flown by every planet taking detailed photos and collecting huge amounts of data. We've even sent rovers to Mars to sample the soil and look for life. All this new information has left us with one big takeaway: Compared with the other worlds around us, Earth is an incredibly active, dynamic planet!

The activity on Earth starts high above the surface with the atmosphere that surrounds it. Earth's atmosphere has a totally different makeup from that of any other planet that we know of. What we call "air" is a complex mixture of gases that are in constant motion. The winds that swirl around the planet are driven by a combination of Earth's rotation as well as the heating and cooling caused by sunlight striking the surface. This process is called convection. Sunlight also causes water to evaporate from the surface, forming a gas in the air called water vapor. When the air cools, the water vapor condenses back into a liquid and eventually falls back to Earth as precipitation. This never-ending process is called the water cycle, and as far as we can tell, Earth is the only planet where it is currently happening.

Once you get down on the surface of Earth, there is even more activity. Water flowing over the land removes soil and broken pieces of rock from one place and deposits them in another, shaping the land in the process. Astronomers believe that Mars used to have water flowing on its surface, too, but from what we can tell, that all happened in the distant past.

WATER, WATER EVERYWHERE!

One of the most important ways Earth stands out from the other planets of our solar system is that we have a natural range of temperatures allowing water to exist in three different states: solid ice, liquid water, and gaseous water vapor. The ability of water to easily change from one form to another is what drives the water cycle, as liquid water evaporates from the surface and falls back to Earth as precipitation. Since water can also exist as a solid, huge quantities of freshwater are stored as ice in mountain glaciers and in the polar ice caps. Unfortunately, because the average surface temperature of the Earth has been steadily rising over the past few decades, many of the world's glaciers are melting at faster and faster rates. This newly melted ice adds liquid water to the oceans, causing sea levels to rise—a big problem for people living near the coasts.

Glaciers often end at the ocean, where they break apart, forming icebergs.

As global temperatures rise, increased evaporation will increase the size and intensity of storms such as Hurricane Katrina, seen here from space in August 2005.

Finally, some of the most impressive activity takes place deep below Earth's surface. Forces working inside the planet cause earthquakes to shake the ground and volcanoes to erupt. Over millions of years, these same forces have moved continents, producing huge mountain chains, some of which are still growing today. Yes, our Earth sure is a dynamic place, and understanding how these different processes work on the planet is the first step in understanding the rocks and minerals we have here!

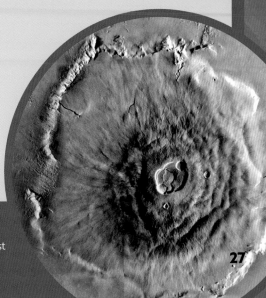

Bet You Didn't Know!

The largest known volcano in the solar system is not found on Earth; it's located on Mars. No longer active, Olympus Mons (pictured) has a total volume that is about 100 times greater than that of Mauna Loa, the largest volcano on Earth. The Martian mountain is also almost three times higher than Mount Everest, Earth's highest peak.

27

GETTING THE
INSIDE STORY

At the beginning of the 19th century, most scientists thought that Earth was simply a solid mass of rock. All this changed when new technology allowed them to track the way that earthquake waves move through the planet. One type of wave can travel only through solids and stops moving when it hits liquids. Other types of waves bend when the density of the rocks that they are moving through suddenly changes. Using this information, they discovered that Earth is really made of several different layers. Let's look at how geologists break them down.

WHERE DID ALL THE **LAYERS** COME FROM?

Most geologists believe that the layered structure of our planet goes back to a time when Earth was still very young and being blasted by all sorts of cosmic debris. The heat generated by all those impacts, combined with the compression caused by gravity pulling the planet together, caused Earth to partially melt. This allowed denser material to sink to the center of the planet and less dense material to float up toward the surface ... just like whipped cream on top of a cup of cocoa!

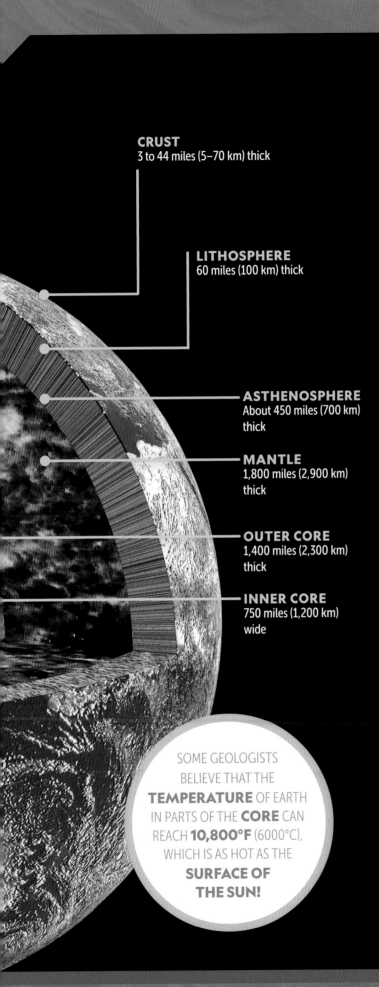

CRUST
3 to 44 miles (5–70 km) thick

LITHOSPHERE
60 miles (100 km) thick

ASTHENOSPHERE
About 450 miles (700 km) thick

MANTLE
1,800 miles (2,900 km) thick

OUTER CORE
1,400 miles (2,300 km) thick

INNER CORE
750 miles (1,200 km) wide

SOME GEOLOGISTS BELIEVE THAT THE **TEMPERATURE** OF EARTH IN PARTS OF THE **CORE** CAN REACH **10,800°F** (6000°C), WHICH IS AS HOT AS THE **SURFACE OF THE SUN!**

CRUST: The outermost layer of Earth is called the crust. It's like a thin skin of brittle rock on the surface, making up only about one percent of the planet. The crust under the ocean is different than the crust under the continents. Oceanic crust is thinner and denser than continental crust and is made up of dark-colored rocks such as basalt (p. 90). Continental crust is thicker and is made up of lighter-colored rocks such as granite (pp. 78–79).

LITHOSPHERE: The lithosphere of Earth is made up of the crust and the uppermost part of the mantle and is composed of solid, brittle rock. Geologists have demonstrated that the lithosphere is divided into several dozen separate chunks called tectonic plates that ride on top of the softer rocks of the asthenosphere below.

ASTHENOSPHERE: Directly below the lithosphere, this layer in the upper mantle is made of dense rock that is not totally solid. Instead, because of extreme heat and pressure, the rock in the upper mantle can slowly flow— almost like toothpaste, but much, much slower—causing the lithospheric plates above to move around.

MANTLE: Earth's mantle makes up a little more than two-thirds of the planet. Though the rocks here are super dense and mostly solid, this layer's high heat and pressure cause the rock in the asthenosphere above it to flow. As you get deeper into the lower part of the mantle, the pressure becomes so great that the rocks become super stiff and lose their ability to flow.

OUTER CORE: Earth's outer core is located below the mantle. It is a liquid layer composed mostly of the metals iron and nickel and some sulfur. Geologists know that it is liquid because certain types of earthquake waves can't penetrate it. Currents moving through this molten metallic layer are thought to generate the magnetic field that surrounds Earth.

INNER CORE: Lying at the very center of Earth is the inner core, believed to be a large, roughly spherical chunk of solid iron and nickel. Scientists had figured out by the start of the 20th century that Earth's outer core is liquid. But the fact that the inner core is solid wasn't discovered until the 1930s, by a Danish scientist named Inge Lehmann. While reviewing data from a large earthquake that happened in 1929, she noticed that certain types of earthquake waves were bouncing off a solid mass at the center of the planet.

SHAKING
ALL OVER

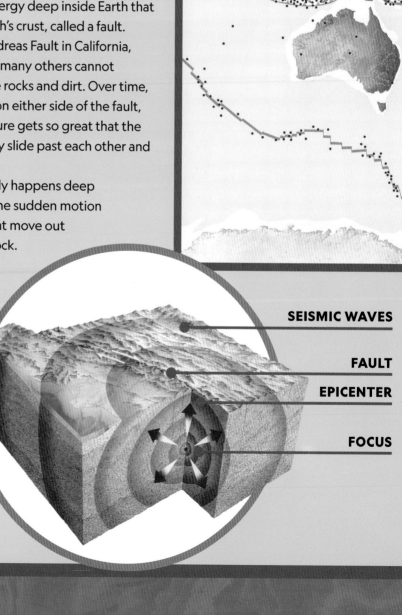

Have you ever experienced an earthquake? Many people have and they don't even realize it! More than a million earthquakes happen around the world each year, but most are too small for us to feel. Earthquakes happen when there is a sudden shift in the crust below the surface. They can be caused by different things, including the eruption of a nearby volcano or the collapse of an underground cave. Most often, though, earthquakes happen because of a sudden release of energy deep inside Earth that causes movement along a crack in the Earth's crust, called a fault.

While some large faults, like the San Andreas Fault in California, U.S.A., can be seen on the planet's surface, many others cannot because the cracks have filled in with loose rocks and dirt. Over time, forces inside Earth push against the rocks on either side of the fault, building up pressure. Eventually the pressure gets so great that the rocks on the two sides of the fault suddenly slide past each other and bingo—earthquake!

The actual movement on the fault usually happens deep underground at a place called the focus. The sudden motion along the fault sets off waves of energy that move out in all directions through the surrounding rock. These earthquake waves are recorded at different locations around the world and, based on their arrival times, scientists can pinpoint the epicenter of the earthquake, which is the point on the ground surface directly above the focus.

Seismologists (the scientists who study earthquakes) use a device called a seismograph to measure the strength of an earthquake. A traditional seismograph has a drum holding

SEISMIC WAVES

FAULT

EPICENTER

FOCUS

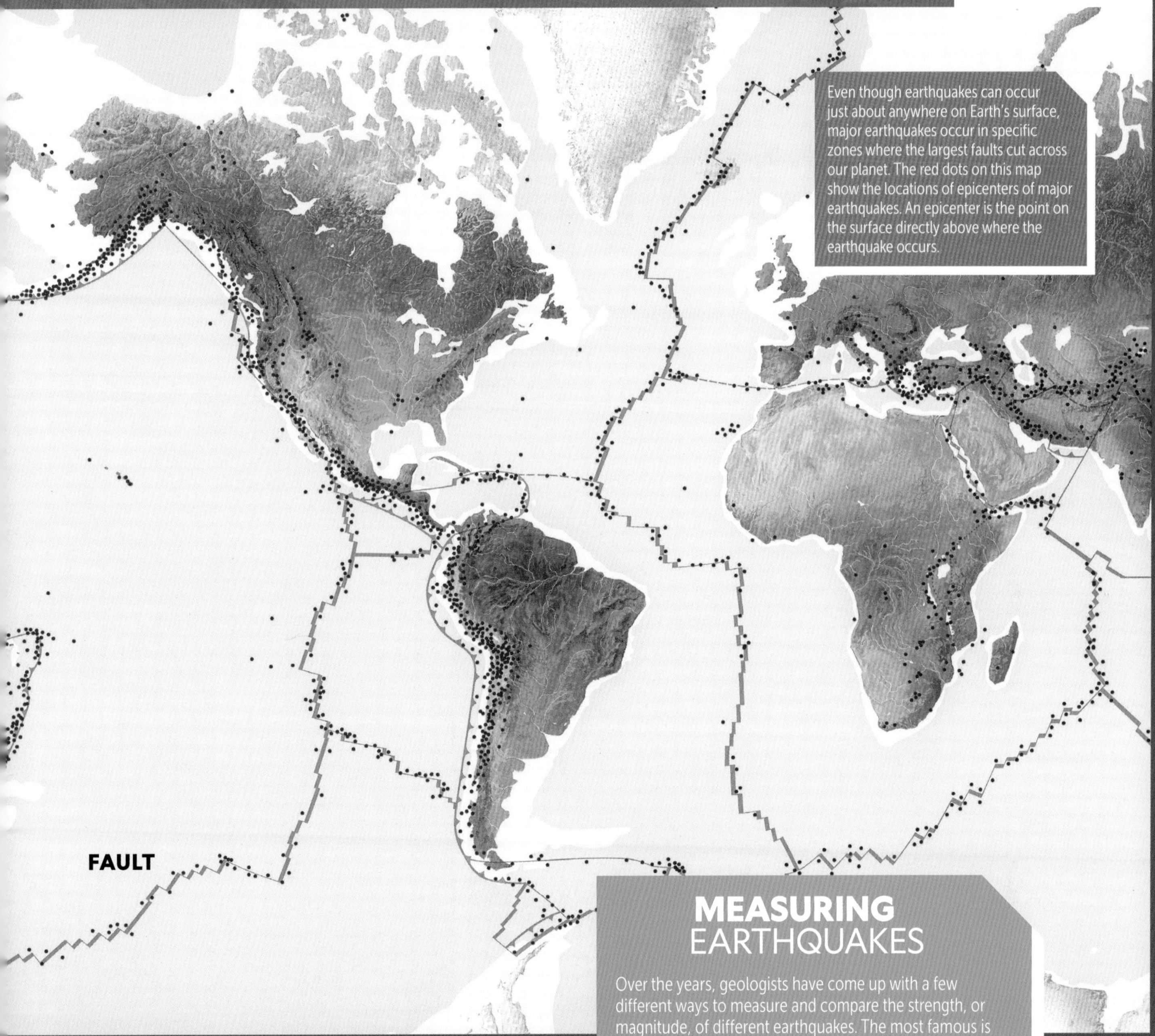

Even though earthquakes can occur just about anywhere on Earth's surface, major earthquakes occur in specific zones where the largest faults cut across our planet. The red dots on this map show the locations of epicenters of major earthquakes. An epicenter is the point on the surface directly above where the earthquake occurs.

FAULT

MEASURING EARTHQUAKES

Over the years, geologists have come up with a few different ways to measure and compare the strength, or magnitude, of different earthquakes. The most famous is called the Richter scale, developed by seismologists Charles Richter and Beno Gutenberg in the early 1930s. These days most geologists prefer to use the moment-magnitude scale, or MMS, which uses seismograph readings to measure both the size of the area where the earthquake happened, as well as the energy that was released. For each number increase on the MMS, the amount of energy released by an earthquake increases about 30 times. In theory, there is no upper limit to the scale. Most earthquakes rank less than a 5, and fewer than 10 earthquakes a year rank more than a 7—these are incredibly destructive. Anything over an 8 is so large that it would pretty much wipe out most of the buildings in an area.

a roll of paper that slowly rotates. A pen that is on a weight is suspended above it. As the ground shakes under the seismograph, the pen does not move, but the ground underneath it does. That causes the pen to draw a line on the paper, recording Earth's movement during the earthquake event. This recording is called a seismogram. The seismogram shows the pattern of waves created by the earthquake: The more powerful an earthquake is, the larger the lines are on the seismogram.

THE WORLD'S **LARGEST MAGNITUDE** EARTHQUAKE WAS **9.5 ON THE MMS,** RECORDED ON **MAY 22, 1960,** NEAR VALDIVIA, IN **SOUTHERN CHILE.**

DO YOU **KNOW** YOUR **FAULTS?**

THE HIMALAYA

Earthquakes happen whenever there is a sudden movement along a geologic fault. A fault is a crack in the Earth's crust along which the rocks have been moved. Sometimes, the forces that act on faults are compressional, which means that rocks are being pushed together. Other times the forces are tensional, as rocks are pulled apart. Rocks can also experience shearing forces, when they are pushed past each other from two different directions. As you might expect, the type of forces that act on a fault control the way the fault looks and moves. Here's a simple guide to the three most common types of faults.

THE EAST AFRICAN RIFT SYSTEM

1

NORMAL FAULT

Normal faults are caused by tensional forces in Earth's crust. As rocks are slowly pulled apart, cracks form. If the rocks on one side of the crack slide down, a normal fault is born. One place where you can find lots of normal faults happening today is in the East African Rift System, a zone more than 4,000 miles (6,400 km) long where forces deep inside Earth are slowly ripping the African continent apart.

GEOLOGISTS BELIEVE THAT THE **SAN ANDREAS FAULT SYSTEM** IN CALIFORNIA IS ABOUT **800 MILES** (1,300 KM) LONG AND HAS BEEN **ACTIVE** FOR BETWEEN **15 AND 20 MILLION YEARS.**

THE SAN ANDREAS FAULT

3 STRIKE-SLIP FAULT

Unlike normal and reverse faults, where the relative motion is up and down, strike-slip faults are caused by shear forces, as one piece of the crust slides past another. You can often see the effects of the motion of strike-slip faults when things like fences and road-ways have been moved sideways. One of the most famous strike-slip faults is the San Andreas Fault system in California, U.S.A., where the piece of the crust under the Pacific Ocean is sliding past the piece of crust under North America.

2

REVERSE FAULT

Reverse faults are sometimes called thrust faults because they are caused by compressional forces in Earth's crust. When rocks are pushed together, they have nowhere to go but up! When one block of rock sides up and over another, a reverse fault is created. One place where you can see thrust faults happening today is in the Himalaya, where India is being pushed up into the rest of the Asian continent.

THE LONG AND **THE SHORT** OF IT

Faults may be very straight with almost no bends in them, or they can be somewhat jagged with lots of twists and turns. Faults also vary in their length. Some might be only a few feet long while others, like the East African Rift Zone, extend for thousands of miles and can be as wide as 40 miles (64 km). Finally, some faults are quite shallow and are found in rocks right near the surface, while others can extend for many miles down into Earth.

THAR SHE BLOWS!

There is nothing as spectacular or terrifying as when a volcano starts to blast superhot material into the air. Trying to predict exactly when a volcano is going to erupt is a lot like predicting the weather: We can get it right most of the time, but sometimes an eruption can surprise us and then, *boom!*

So, what exactly is a volcano? Well, to put it simply, it's an opening in Earth's crust through which material from inside the planet can make its way to the surface. Volcanoes are important because they add new rocks and minerals to the crust. Without them, we wouldn't have tropical paradises like the U.S. Hawaiian Islands!

Volcanoes get their start when hot molten rock called magma forms pools called magma chambers below the surface. Geologists aren't 100 percent sure where this magma originates. Some of it may be produced in the lower crust, and some of it may come from "plumes" that are heated by the core and rise up through the mantle. Eventually, pressure begins to build in the magma chamber and the magma starts to work its way up through an opening called a vent. Once the magma reaches the surface and flows out over the ground, it is called lava. Some types of lava, like the stuff that comes out of the Hawaiian volcanoes, flows easily and acts almost like a red-hot river. Other volcanoes have magma that is super thick and flows really slowly. It's these volcanoes that can be extremely dangerous, because their vents can get plugged up by the magma, allowing pressure to build up and causing them to explode.

Not all volcanoes erupt just lava. Some also blast out thick clouds of tiny superhot rock fragments called volcanic ash. Often these ash clouds can rise tens of thousands of feet into the air, causing major problems for airplanes flying overhead and for people living hundreds of miles away in places where the ash finally lands. Volcanoes also release toxic gases, water vapor, and tiny particles called aerosols that can stay up in the atmosphere for months at a time and are carried around the world by global wind currents.

ASH CLOUD

CRATER

LAVA FLOW

SIDE VENT

LAYERS OF ASH

MAIN VENT

ROCK LAYERS OF THE EARTH'S CRUST

MAGMA CHAMBER

The hot springs and geysers of Yellowstone National Park, in the western United States, are being heated by a gigantic pool of magma deep underground. Around 640,000 years ago, this magma produced a huge volcanic eruption, making Yellowstone one of the largest supervolcanoes in the world.

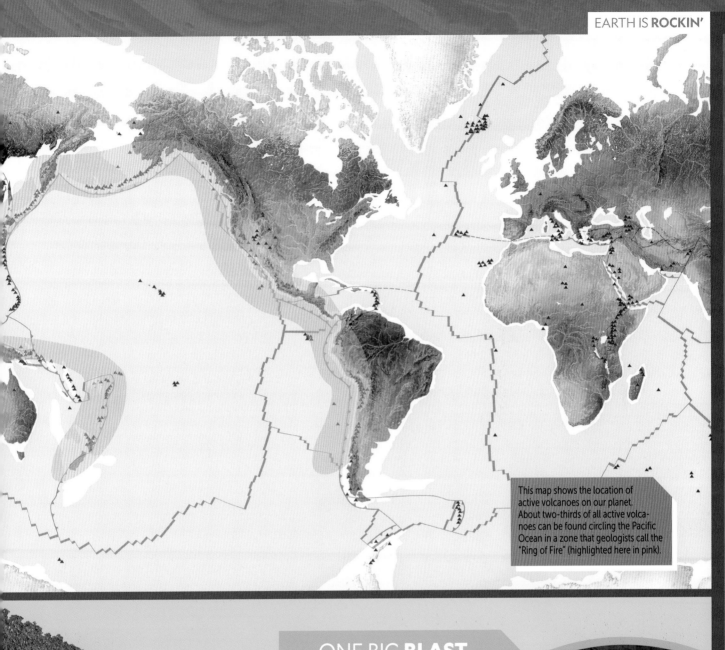

This map shows the location of active volcanoes on our planet. About two-thirds of all active volcanoes can be found circling the Pacific Ocean in a zone that geologists call the "Ring of Fire" (highlighted here in pink).

ONE BIG **BLAST**

The largest volcanic eruption ever recorded happened on the Indonesian island of Sumbawa on April 10, 1815. That's when the entire top of Mount Tambora literally blew off, removing almost 5,000 feet (1,524 m) of rock and soil from the top of the mountain. Geologists estimate that the eruption blasted more than 30 cubic miles (125 cubic km) of volcanic material into the air, including tiny particles of ash and sulfur that hung around in the atmosphere for months. These particles blocked so much sunlight from reaching Earth's surface that, for almost a year, average global temperatures dropped by as much as 5°F (3°C). Crops failed around the world, causing 80,000 people to die of starvation and disease. In fact, things got so cold in parts of Europe and North America that heavy snow fell in July and August, making 1816 the so-called year without a summer.

GEOLOGISTS CALL ANY TYPE OF MATERIAL THAT COMES OUT OF A VOLCANO **EJECTA.** THIS CAN INCLUDE LIQUID LAVA, TINY PIECES OF VOLCANIC ASH, OR EVEN LARGE ROCKS CALLED **VOLCANIC BOMBS!**

MAJESTIC **MAGMA** MOUNTAINS

You might think that one volcano is pretty much the same as the rest. They all start with magma, look like mountains (or at least hills), and spew some type of ejecta from inside the planet. But that's where the similarities end. Some volcanoes are pointed and tall with steep slopes, while others are broad and kind of flat. There is even one type of volcano that forms inside another volcano! After studying hundreds of different volcanoes that dot our planet, geologists have been able to break them down into four main groups.

COMPOSITE VOLCANO: MOUNT FUJI, JAPAN

When most people hear the word "volcano," composite cones are the type they usually picture. These massive mountains get built up over time as different types of volcanic materials erupt. These include lava flows, volcanic ash, cinders, and even volcanic bombs. Composite volcanoes, also called stratovolcanoes, have steep sides and often grow to be more than 8,000 feet (2,350 m) tall. The magma that feeds composites is thick and flows very slowly. As a result, the magma can sometimes plug the vent before flowing out the volcano's top, which is kind of like putting your thumb on top of a bottle of soda and shaking it. If enough pressure builds up from below, the entire top can blow off the mountain. Some of the most famous volcanoes in the world are composite volcanoes, including the perfectly cone-shaped Mount Fuji in Japan.

SHIELD VOLCANO: MAUNA LOA, HAWAII, U.S.A.

Shield volcanoes are some of the largest volcanoes. They build up over many thousands of years, as layer upon layer of lava flows out of a series of vents and piles up on top of one another, forming layers of rock called basalt (p. 90). Shield volcanoes get their name from the fact that they are broad with a rounded surface and look a little like a gladiator's shield laying flat on the earth. They are found in long chains going across parts of the Pacific Ocean and have built many islands, including the Big Island of Hawaii, where the Mauna Loa volcano is found.

LAVA DOME: MOUNT PELÉE, MARTINIQUE

Technically speaking, lava domes are not usually free-standing volcanoes. Most often, they form inside the top or on the side of a composite cone volcano. Because the magma in these volcanoes is super thick, it has a hard time flowing up through the vent. So when it finally comes out as lava, it pretty much hardens in place. As more magma tries to work its way out of the vent, it pushes the hardened lava above it up, forming a dome of rocks with sharp, jagged edges. Sometimes lava domes will suddenly get a renewed charge of gas from below and erupt violently. In 1902, a lava dome on the side of Mount Pelée on the Caribbean island of Martinique erupted, releasing a superhot burst of ash and gas that raced down the side of the mountain, killing some 30,000 people in the town of St. Pierre—which was located four miles (6.4 km) away.

CINDER CONE: PARÍCUTIN, MEXICO

Cinder cones are relatively small volcanoes that rarely have liquid lava flowing down their sides. Instead, as the name suggests, they are made up almost entirely from small solid particles called cinders that form when liquid lava gets blasted high up into the air under great pressure. As the small blobs of lava quickly cool, they turn solid and fall back down, building a round cone around the vent. Most cinder cone volcanoes erupt for only a few years and then become quiet, but they can grow very quickly. One of the most famous is Paricutin, in the Mexican state of Michoacán. It started erupting in the middle of a farmer's field in 1943 and, by the time it had stopped nine years later, had grown to a height of about 1,390 feet (424 m).

BIG HAPPENINGS UNDER THE SEA

Have you ever gone swimming at an ocean beach? Unless it's a really rocky part of the shoreline, the bottom feels smooth and flat and is usually covered with sand. Long ago, this is pretty much what most people imagined the entire ocean floor was like: smooth, flat, and just covered with lots of sand. Some people did have other ideas, though. Fishermen, for example, sometimes got their nets stuck on big chunks of rock or coral that stuck up from the seafloor. But people thought these occurrences were rare. It really wasn't until the beginning of the 20th century that the picture of the ocean floor started to change, thanks to the invention of submarines and a device called sonar.

After people had the ability to travel deep into the ocean and see for themselves what was going on down there, they were amazed at what they found. The ocean floor wasn't simply a broad, flat plain. It resembled the land surface and had large mountains, deep valleys, and even underwater volcanoes spewing lava and thick clouds of black smoke. The problem is that the ocean is really big, and even with a bunch of subs taking the plunge, scientists could see just a tiny part of what was happening down there. Sonar helped bring things into focus.

Sonar uses sound waves to measure how far away an object is. During World War II, navies on both sides of the conflict used it to locate and track enemy submarines. It works in a way similar to the echolocation system that bats and dolphins use for finding food. A device called a transducer sends out pulses of sound waves, which bounce off objects in the water. The operator can then figure out how far away the object is based on the speed of sound in water and how long it takes for the reflected wave to return to the transducer. A clever geologist named Harry Hess,

Found in volcanically active areas under the sea, hydrothermal vents release superhot mineral-rich water from inside the crust.

WHILE THE OLDEST ROCKS ON **EARTH'S CONTINENTS** ARE MORE THAN **FOUR BILLION YEARS OLD,** THE OLDEST PIECES OF **OCEANIC CRUST** ARE ONLY AROUND **200 MILLION** YEARS OLD!

Bet You Didn't Know!

The Mid-Ocean Ridge system is the longest mountain chain in the world, winding for more than 40,000 miles (65,000 km) around the globe. It is the place where new oceanic crust is created. Most of the mountains found on the ridge are deep underwater, but in some places, like those that form the island of Iceland, they rise above the surface of the sea.

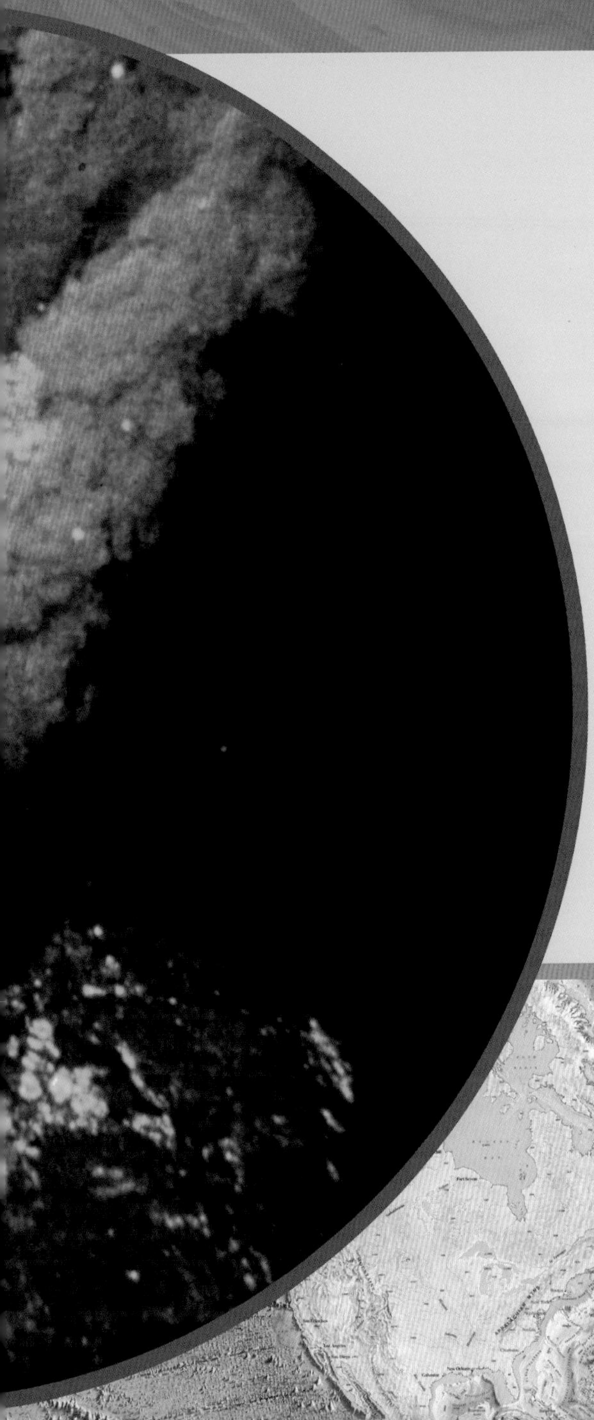

who happened to be in the U.S. Navy during the war, realized that sonar could also be used to "see" what was going on at the bottom of the ocean. He started taking sonar readings as his ship sailed back and forth across the Pacific Ocean. That's when geologists began to realize just how active the crust under the ocean is and why volcanoes and earthquakes happen in the places they do.

THE MAPS OF
MARIE THARP

One of the most important developments that helped geologists understand crustal motions under the ocean came in 1977 with the publication of the first accurate map of the entire ocean floor. This map was the product of 30 years of work by geologists Marie Tharp and Bruce Heezen. Heezen sailed around the world using sonar to take depth readings of the seafloor, and Tharp turned this raw data into detailed drawings showing what this seafloor looked like. In doing so, Tharp discovered features including a long chain of underwater mountains (called the Mid-Ocean Ridge) running around the entire planet and large valleys in the seafloor (called submarine trenches) at the edge of some of the continents. Both features would allow other geologists to work out the theory of plate tectonics, which says that Earth's rigid outer shell (the lithosphere) is divided into plates that move around over the mantle. Plate tectonics is one of the most important concepts in geology today.

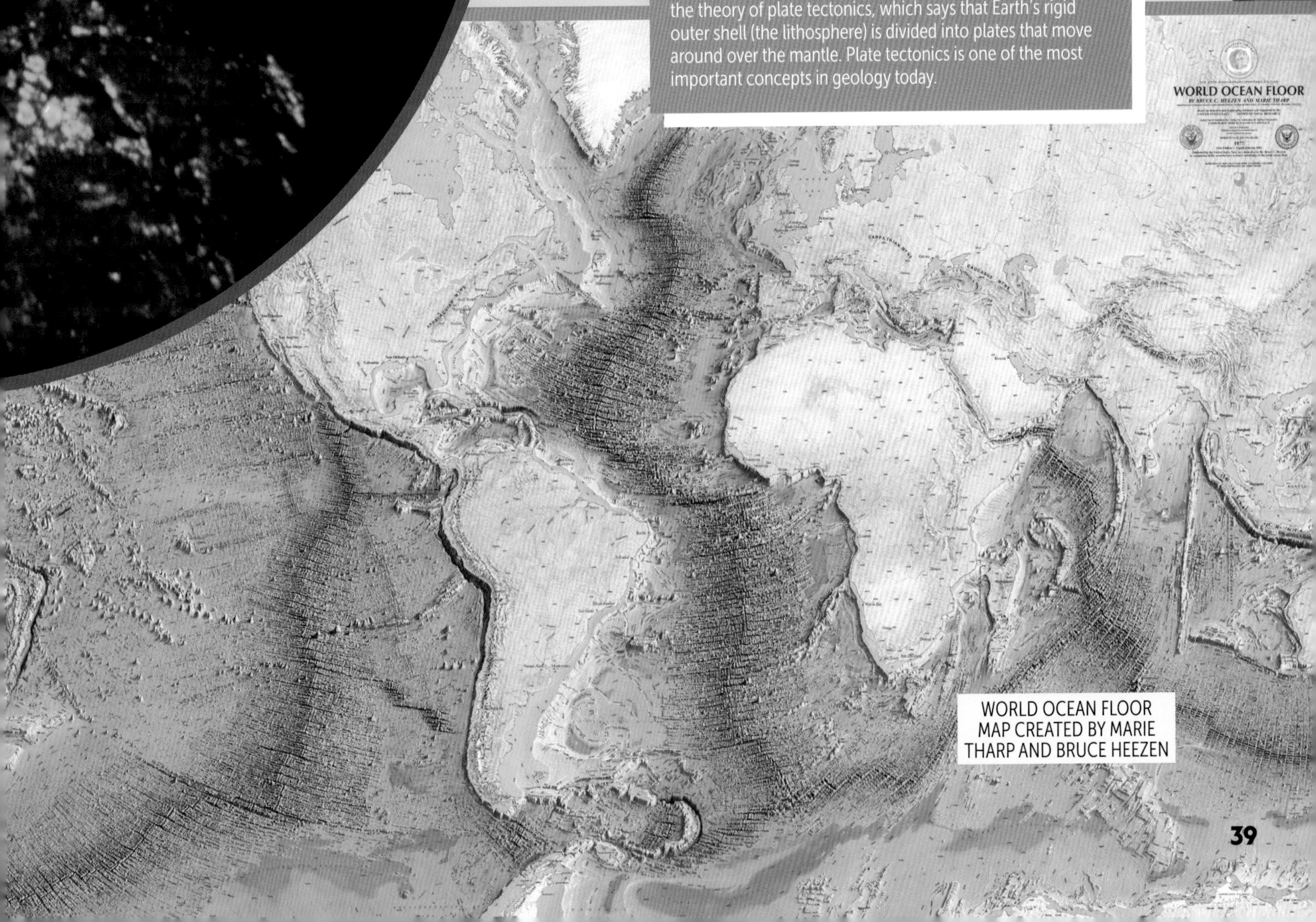

WORLD OCEAN FLOOR
MAP CREATED BY MARIE
THARP AND BRUCE HEEZEN

A **RIP** IN THE OCEAN **FLOOR**

People love record breakers—especially on an epic scale.

The highest mountain on Earth? It's Mount Everest, in the Himalaya. The largest island in the world (not counting the continental landmasses of Australia and Antarctica)? That would be Greenland. And the deepest place in the ocean? The Mariana Trench, coming in at a little more than 36,000 feet (11,000 m) below sea level.

Located in the western part of the Pacific Ocean about 124 miles (200 km) east of the Mariana Islands (from which it got its name), the Mariana Trench is basically a large rip in the bottom of the ocean floor that geologists call a subduction zone. Subduction zones form when a section of oceanic crust is jammed back into the Earth to eventually become part of the mantle again. These zones can be found along the edge of several continents.

When we say that the Mariana Trench is a big rip, we're not kidding. The crescent-shaped chasm is a little over 1,500 miles (2,500 km) long and averages about 43 miles (69 km) wide. While there are a number of deep spots along the length of the trench, the lowest point is a steep-sided valley called the Challenger Deep, found southwest of the island of Guam.

Scientists are not certain exactly how deep the Challenger Deep is. Even with modern technology, getting precise measurements that far down is difficult. The first attempt to measure the depth of the trench happened in 1875, when the crew of the British research ship

The Mariana Trench isn't the only deep hole in the ocean floor. So far scientists have identified more than 20 different submarine trenches, many located along the edges of continents around the rim of the Pacific Ocean.

SEABED GARDENS IN
A SEAMOUNT OF THE
MARIANA TRENCH

EXPLORING THE
CHALLENGER DEEP

Is it possible to dive to the bottom of the Mariana Trench? Not only is it possible, it's been done three times!

In January 1960, U.S. Navy officer Don Walsh and Swiss engineer Jacques Piccard took a crewed submersible called a bathyscaphe down into the Mariana Trench. Their dive in the *Trieste* (right) took them about five hours to reach the seafloor and proved that a vessel could survive such a deep dive.

In March 2012, explorer and filmmaker James Cameron (below) piloted a one-person submersible he codesigned named *DEEPSEA CHALLENGER.* Cameron explored the seafloor for two and a half hours, collecting a sediment sample and the first video footage at this crushing depth.

On April 28, 2019, ocean explorer Victor Vescovo dived in the submersible *Limiting Factor.* Vescovo made five dives into the Mariana Trench, recording video footage showing strange life-forms in the deepest parts of the ocean.

All three dived to a depth of 35,810 feet (10,915 m), the deepest place ever measured in our oceans.

THE **DEEPEST PART** OF THE **MARIANA TRENCH** IS **SO DEEP** THAT IT COULD FIT ALL OF **MOUNT EVEREST** AND STILL HAVE ALMOST **7,000 FEET** (2,130 M) OF WATER **ABOVE IT!**

H.M.S. *Challenger* used a long weighted rope and came up with a reading of 26,850 feet (8,184 m) at the southern end of the trench. Over the years, many other measurements at other locations along the trench have been made using more sophisticated sonar equipment, and the measurement has steadily increased. Time will tell if scientists discover a deeper spot on the planet, but for now, the Challenger Deep in the Mariana Trench is the deepest known rip in the world.

DRIFTING CONTINENTS AND SHIFTING PLATES

Do you like to do jigsaw puzzles? If you take a close look at a globe, you'll see that many of the continents look like they can fit together, almost like pieces of a giant jigsaw puzzle. Is this just a coincidence, or is there a reason that the continents are shaped this way? This same question puzzled geologists going back to the 17th century, but it wasn't until the early part of the 20th century that they found the answer. The reason the continents look like they can fit together is because about 250 million years ago they were joined together!

To understand what's going on, we need to do a little geo-detective work. Look at the map on pages 30 and 31 showing the location of major earthquakes in the world and compare it with the map on pages 34 and 35 showing the location of active volcanoes. Notice any similarities? You bet! Most of the places that have active volcanoes also get lots of big earthquakes! This makes sense because earthquakes happen as a result of movement along cracks in the crust called faults, and it's also cracks in the crust that allow magma to rise to the surface to make volcanoes. Do you think these cracks are related?

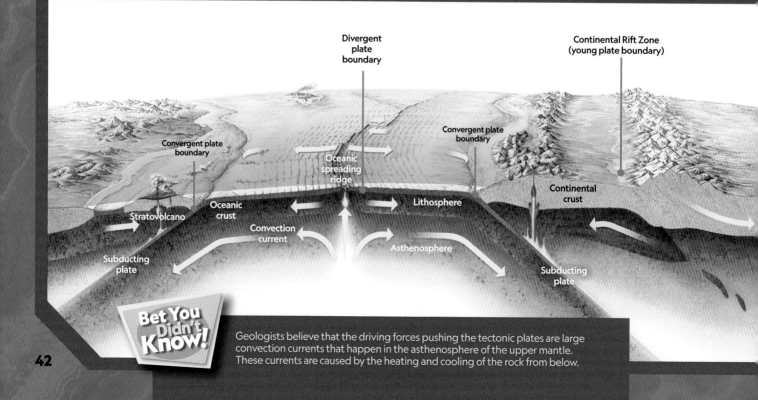

Divergent plate boundary

Continental Rift Zone (young plate boundary)

Convergent plate boundary

Convergent plate boundary

Oceanic spreading ridge

Continental crust

Stratovolcano

Oceanic crust

Lithosphere

Convection current

Asthenosphere

Subducting plate

Subducting plate

Bet You Didn't Know!

Geologists believe that the driving forces pushing the tectonic plates are large convection currents that happen in the asthenosphere of the upper mantle. These currents are caused by the heating and cooling of the rock from below.

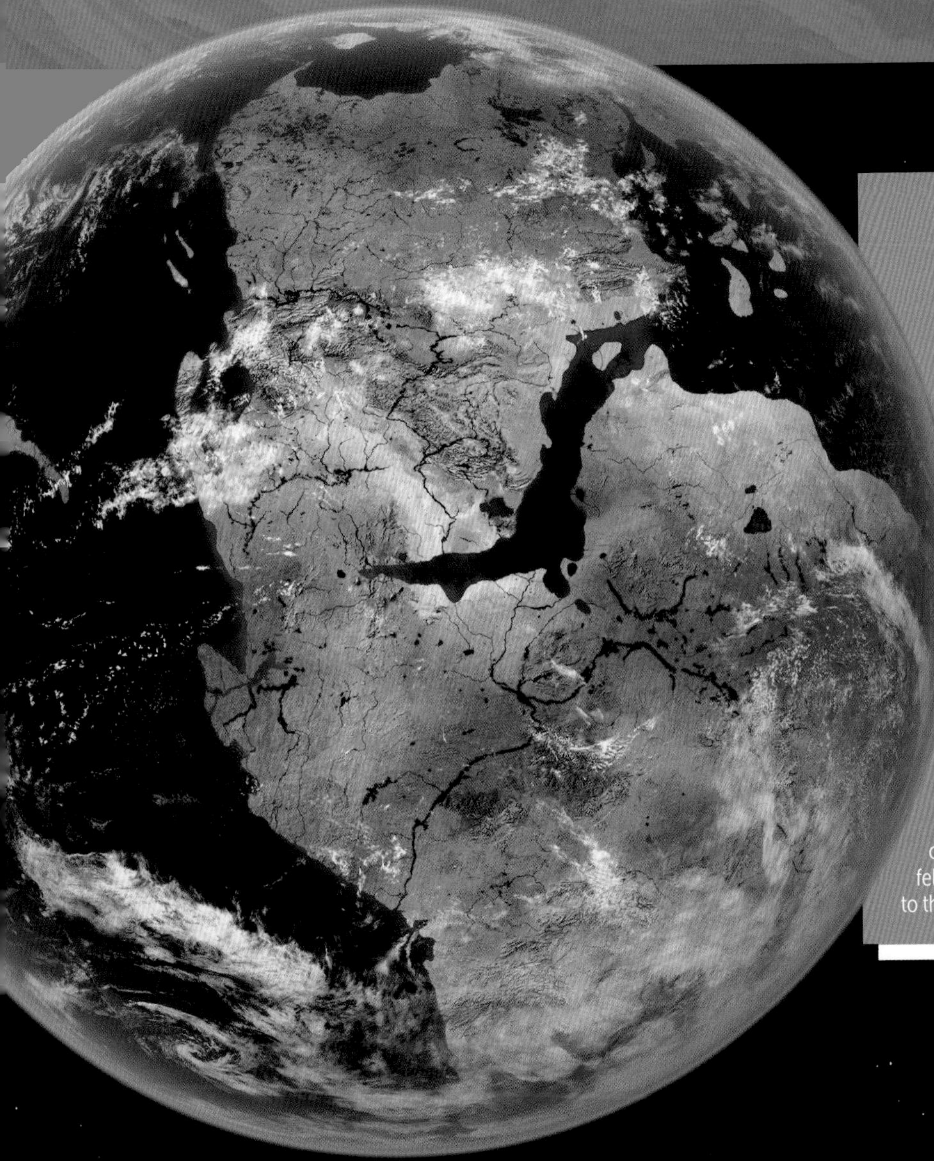

THE MAN WHO **MOVED CONTINENTS**

Alfred Wegener moved continents! While he was not the first scientist to suggest that continents once fit together, his research opened the door for other scientists who proved they did. Wegener didn't just base his idea of continental drift—the movement of continents—on the shape of the continents. He conducted a detailed analysis of rock types and fossils on the different landmasses to show how they had been joined.

In 1915, he published a book called *The Origin of Continents and Oceans*, in which he explained his theory and proposed that all of Earth's landmasses had once made one giant supercontinent, which he called Pangaea. At first, few scientists took his ideas seriously, but over the next 20 years, other scientists would demonstrate just how the continents could move. Little by little, the pieces fell into place and, in the end, Wegener's ideas led to the theory of plate tectonics.

DIVERGENT PLATE BOUNDARIES AREN'T JUST FOUND ON THE OCEAN FLOOR. THE **EAST AFRICAN RIFT VALLEY** IS A DIVERGENT BOUNDARY ON LAND WHERE THE CONTINENT IS BEING **RIPPED APART.**

The answer can be found on the map of the ocean floor that Marie Tharp drew (pp. 38–39). If you follow the lines created by the Mid-Ocean Ridge and the submarine trenches, you can actually see the cracks!

It turns out that Earth's crust is not one solid mass but is instead made up of several dozen individual chunks called tectonic plates, with cracks between them called plate boundaries. Tectonic plates don't just sit there—they move slowly in different directions. Along the Mid-Ocean Ridge, the plates move away from each other. These are called divergent boundaries. Convergent boundaries can be found at the subduction zones along the trenches where the plates crunch together and the oceanic crust gets pushed back into Earth. You can also have places where one plate slides past another. These are called transform boundaries. When you put all these pieces together, you get one really big idea. It's called plate tectonics!

A GORGE IN THE EAST AFRICAN RIFT AT ENGARUKA, TANZANIA

ALL CRACKED UP!

Just like the cracked shell of a hard-boiled egg, the outer shell of Earth is broken into a few dozen solid chunks called lithospheric plates, which form the crust and uppermost mantle. The plates of the lithosphere are resting on the semisolid rock of the asthenosphere below, through which large, slow-moving convection currents flow. These convection currents are what move the plates. Magma that works its way up to the surface along the Mid-Ocean Ridge pushes the plates apart and makes new crust as it cools. At subduction zones, old, dense oceanic crust gets forced back down under the continental plates to return to the mantle. The seven largest plates are called major tectonic plates, and they cover almost 95 percent of the Earth's surface.

3 EURASIAN PLATE

2 NORTH AMERICAN PLATE

12 JUAN DE FUCA PLATE

AFRICAN PLATE 4

10 CARIBBEAN PLATE

11 COCOS PLATE

1 PACIFIC PLATE

8 SOMALI PLATE

7 SOUTH AMERICAN PLATE

9 NASCA PLATE

5 ANTARCTIC PLATE

1 PACIFIC PLATE

At more than 39,900,000 square miles (103,300,000 sq km), this is by far the largest tectonic plate. Most of the Pacific plate is made up of oceanic crust, except for some small slices of continental crust around New Zealand and Southern California. The plate is surrounded by the Ring of Fire, home to more than two-thirds of the world's active volcanoes.

2 NORTH AMERICAN PLATE

Coming in second, with an area of about 29,300,000 square miles (75,900,000 sq km), this plate has large areas of both continental and oceanic crust. The eastern edge begins in the middle of the Atlantic Ocean, at the Mid-Atlantic Ridge.

3 EURASIAN PLATE

Including most of Europe and Asia, this plate has an area of about 26,100,000 square miles (67,800,000 sq km), making it the third largest. The western edge of the plate starts at the Mid-Atlantic Ridge and continues east to include the islands of Japan and Malaysia.

4 AFRICAN PLATE

Coming in at number four, at around 23,600,000 square miles (61,300,000 sq km), is the African plate. Like the Eurasian plate to the north, it starts in the west at the Mid-Atlantic Ridge. On the east, the African plate ends at the East African Rift Zone, where it is separating from the Somali plate. Over time, geologists believe that the African landmass will split into two pieces!

5 ANTARCTIC PLATE

Just slightly smaller than the African plate, at about 23,500,000 square miles (60,900,000 sq km), is the Antarctic plate. This plate is the fifth largest and basically covers the entire southern part of Earth, including both the continent of Antarctica as well as large parts of the crust under the waters surrounding the continent.

6 INDO-AUSTRALIAN PLATE

The Indian and Australian plate makes up the sixth largest in area, covering almost 22,740,000 square miles (58,900,000 sq km). Many geologists think that this plate should now be classified as two separate plates, because they appear to have split apart about three million years ago.

7 SOUTH AMERICAN PLATE

At only about 16,800,000 square miles (43,600,000 sq km), this is by far the smallest of the major tectonic plates. Like its North American cousin, the South American plate begins at the Mid-Atlantic Ridge to the east and ends abruptly along the edge of the continent to the west, where it crashes into the Nasca plate under the Pacific Ocean.

8 SOMALI PLATE

Covering about 6,448,000 square miles (16,700,000 sq km), the Somali plate is the largest of the minor plates. It includes a big part of the eastern portion of the African continent starting at the East African Rift Zone and extends out into the Indian Ocean, including the island of Madagascar. Over time, a new ocean will form as the Somali plate moves away from the African plate.

9 NASCA PLATE

With an area of roughly 6,000,000 square miles (15,600,000 sq km), this is the second largest of the minor plates. Even though it is relatively small, the Nasca plate plays a big role in the tectonic activity in the eastern Pacific Ocean, pushing up against the South American plate. This action created the Andes Mountains, which are still rising today.

10 CARIBBEAN PLATE

Underlying Central America, the Caribbean Sea, and the island of Hispaniola, this minor plate has a total area of about 1,200,000 square miles (3,300,000 sq km). Squeezed between the North and South American plates, the movement of the Caribbean plate is the cause of many major earthquakes.

11 COCOS PLATE

At a little less than 1,120,000 square miles (2,900,000 sq km), this minor plate is made up of mostly oceanic crust and has a major role in creating earthquakes in Mexico and the countries of Central America, as it slides down under the North American plate.

12 JUAN DE FUCA PLATE

At only 96,500 square miles (250,000 sq km), this plate is officially classified as a "microplate." But it is one of the most tectonically active places on the planet and has triggered numerous earthquakes. It also has produced many active volcanoes that are part of the Ring of Fire.

6
INDIAN PLATE

6
AUSTRALIAN PLATE

ONWARD AND UPWARD

Some people have a passion for climbing mountains. Maybe it's because they like the view from way up top, or maybe it's just because humans feel a sense of triumph over nature when they scale a high peak. So, what exactly are mountains, and how do they get to be so tall?

A mountain is simply a very tall feature on Earth's surface that sticks up high above the surrounding area. There is no official height that a mountain needs to be to be called a mountain, and mountains can even form under the ocean.

There are a few ways that a mountain can form. When lava and ash come spewing out of a volcano, it eventually builds up a cone around the volcanic vent and—presto!—you have a mountain. Classic examples of volcanic mountains are Mount Fuji in Japan and Mount Hood in the United States. Another way to make a mountain is to have some type of hard rock, like granite (pp. 78–79), surrounded by softer rock, like shale (p. 127), exposed on the surface. When the softer rocks wear away, the hard rock is left standing up by itself. Mount Katahdin in Maine, U.S.A., is this type of backward-forming mountain.

To make large groups of megamountains, such as the Alps in Europe or the Himalaya in Asia, you need a little help from plate tectonics. All the major mountain chains on the continents formed as a result of one tectonic plate slamming into another. This can happen in a few ways. The Andes, for example, are located along the western edge of the South American plate and were pushed up as the rocks of the Nasca plate were forced back into the mantle along the edge of the continent. The Himalaya were born from an even more intense collision. Starting around 50 million years ago, the Indo-Australian plate rammed into the bottom of the Eurasian plate—and it's still pushing today. Since both of these plates are made from continental crust, neither is forced down into Earth, so they have nowhere to go but up!

THE TALLEST MOUNTAIN IN THE WORLD

Quick, what's the tallest mountain in the world? Most people would answer Mount Everest ... but they would be wrong! At 29,035 feet (8,850 m) above sea level, Mount Everest, in the Himalaya, is the highest mountain in the world. Technically speaking, the *tallest* mountain is the Mauna Kea volcano, in the U.S. Hawaiian Islands. Mauna Kea rises out of the ocean with its peak topping out at 13,796 feet (4,205 m) above sea level, more than 15,000 feet (4,600 m) lower than Mount Everest. Wait, what?! How can it be taller if it's *lower*?

It all has to do with how you define how tall a mountain is. When geologists speak about a mountain being "tall," they are measuring a mountain from its base to the top of its peak. If you measure Mauna Kea from its base, which is on the bottom of the ocean about 19,700 feet (6,000 m) below sea level, it is more than 33,000 feet (10,000 m) tall—more than 3,000 feet (900 m) "taller" than Mount Everest! Check out the graphic at right.

Mountaineers flock to places like Chamonix, in southeastern France. Chamonix is home to Mont Blanc, the highest mountain in Western Europe.

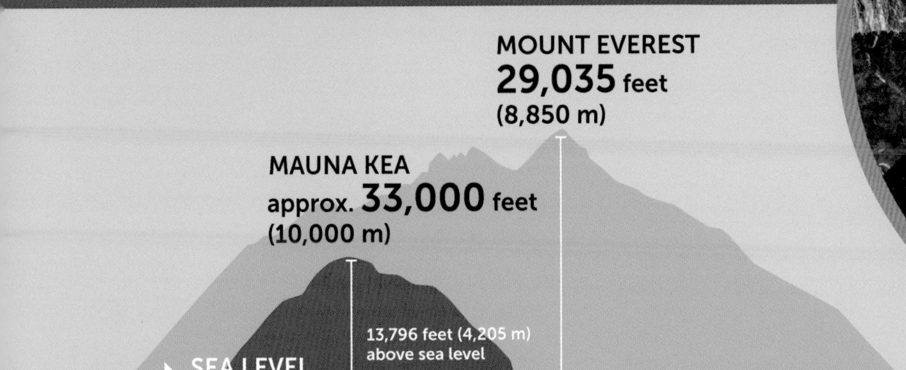

MOUNT EVEREST
29,035 feet
(8,850 m)

MAUNA KEA
approx. **33,000** feet
(10,000 m)

13,796 feet (4,205 m) above sea level

▶ **SEA LEVEL**

19,700 feet (6,000 m) below sea level

MOUNT EVEREST, THE HIGHEST MOUNTAIN ON EARTH, IS **STILL GROWING.** CURRENT ESTIMATES SAY THAT IT'S RISING AT A RATE OF MORE THAN **HALF AN INCH (1 CM) A YEAR!**

One of the first people to explain how mountains formed was Leonardo da Vinci. That's right, the same guy who painted the "Mona Lisa" was also a man of science who loved rocks! Leonardo correctly reasoned that the rocks had originally formed under the sea and were pushed up to make the mountains we see today.

WEATHERING AWAY

Here's a little experiment for you to try: Pick up a small rock and squeeze it in your hand. Can you break it apart? Chances are you couldn't, because most rocks are super hard, and even smashing them with a hammer takes a great deal of effort. So, if rocks are so tough to break, why is there so much sand covering the beaches of the world? After all, beach sand is just tiny pieces of broken-up rock and shells.

It turns out that Mother Nature has ways of breaking up even the hardest rocks. The process is called weathering, and it happens in a few cool ways. And when we say "cool," we aren't kidding! When water freezes into solid ice it expands and takes up more space. That's why when you leave a drink in the freezer a little too long the container swells up and can sometimes even burst. When water seeps into tiny cracks in rocks and then freezes, the same thing happens. The ice expands and, little by little, the tiny cracks get bigger and bigger until the rock splits apart. This type of weathering is called frost wedging, and it happens anywhere that temperatures get below freezing.

Another way that rocks get broken is when they are moved by wind, water, or flowing ice and, in the process, bang into each other. This is called abrasion. Ever get caught in a wave at the beach and get rolled along the bottom? You can really get scraped up. Now think about all

ROCKY CLIFF SHOWING THE EFFECTS OF FROST WEDGING

BIG TEMPERATURE CHANGES CAN ALSO **BREAK APART ROCKS.** WHEN ROCKS EXPAND AND CONTRACT FROM HEATING AND COOLING, THEIR OUTER LAYERS **CRACK AND PEEL OFF** IN A PROCESS CALLED **EXFOLIATION.**

Bet You Didn't Know!

Much of the salt found in seawater comes from rocks and minerals on the land. When rainwater runs over the rocks, it chemically reacts with the minerals, causing them to dissolve. The water with the dissolved minerals in it eventually collects in the ocean.

TREE ROOTS, LICHENS, AND **MOSS,** OH MY!

Even though we humans might find it difficult to break apart rocks, there are plenty of living things that work hard at it every day. Take lichens, for instance. These simple organisms, which are actually part algae and part fungus, literally eat rocks by releasing chemicals that dissolve the minerals in them. You can frequently see them covering rocks, especially up in the mountains where they help to create soil for plants to grow in.

Plants can also do a number on rocks. Like lichens, there are a few varieties of mosses that live on rocks and get the minerals they need directly from the rock. The champion rock wreckers, though, are trees. Tree roots in search of water and nutrients will grow right down into cracks in rocks and, in the process, split the rocks apart. Scientists call this biological weathering, but to a tree, it's just their way of making a living!

YELLOW LICHENS

those rocks at the beach getting pounded by waves 24 hours a day. It's no wonder there is so much sand at the beach!

Both frost wedging and abrasion are forms of mechanical weathering because they physically break the rock apart. Water can also weather rocks in a much quieter way. When water flows over rocks, it can react chemically with the minerals in them, causing the minerals to dissolve. This same process happens when you put salt in a glass of water and mix it up. The salt seems to disappear, but if you taste the water, you know it's in there. When water or other substances dissolve minerals, it's called chemical weathering.

THE WORLD DOWN BELOW

When people hear the word "cave," they often think of spiders, snakes, and, of course, bats! While these creatures do live in caves, to a geologist, the coolest parts of a cave are the strange rock formations called speleothems often found inside.

A cave is any large covered space found in or between rocks. Caves form in several different ways. Sea caves get carved out of cliffs along the shoreline due to the constant pounding of waves. Eolian caves are usually found in desert areas where they have been blasted out of the rock by sand carried by strong winds. There are even caves that form from volcanic activity when the outside of a lava flow quickly cools and forms long hollow tubes. But by far the largest and most common caves in the world form when water wears away softer rocks like limestone (pp. 120–121), dolomite (pp. 194–195), and gypsum (pp. 200–201) below Earth's surface.

Limestone caves, which are known as karst systems, can be found all over the world. Some, like Mammoth Cave in the U.S. state of Kentucky, are huge, stretching for hundreds of miles. Karst caves usually form when water seeps into the ground through cracks in some type of harder cap rock like shale (p. 127) or sandstone (p. 127). When the water reaches the softer limestone, it begins to dissolve it away, creating a small hole. Over time, as more water flows into the space, the hole gets larger and larger until a cave is born.

Large limestone cave systems take thousands or even millions of years to form and are lots of fun for the people called spelunkers who explore them. Water does more than create caves—it can also deposit new rock as it drips down from the cave ceiling creating speleothems. Stalactites hang from the cave roof and look like stone icicles. Stalagmites look like upside-down ice-cream cones and grow up from the floor underneath them when water deposits minerals on the cave floor. When the two connect, you get stone columns. There are even speleothems that look like soda straws and ribbons. With all the geologic activity happening in them, caves can truly be a wonder down under!

UNIQUE CREATURES OF THE UNDERWORLD

What do cave salamanders, eyeless cave fish, and cave crayfish all have in common? Sure, they all live in a cave, but to be specific, they all live in the same cave—Mammoth Cave in Kentucky. All three of these critters, along with a bunch of insects, spiders, and millipedes, call Mammoth Cave home. Scientists have a special name for creatures that live their entire lives in caves. They're called troglobites, and many have some unique adaptations. Eyeless cave fish, for example, have no eyes because they don't need them in an environment that is totally dark. Cave crayfish and cave shrimp are almost see-through because, in the dark, they don't need pigment in their exoskeletons.

So why aren't bats on this list? Since bats live only part of their lives in caves, they are not true troglobites. Scientists have a cool name for them, too, though. They are called trogloxenes, a group that also includes some types of snakes and spiders!

MAMMOTH CAVE CRAYFISH

A spelunker crawls through Mammoth Cave.

MAMMOTH CAVE IN KENTUCKY IS THE **LONGEST CAVE SYSTEM IN THE WORLD,** WITH OVER **400 MILES** (644 KM) OF **PASSAGES MAPPED** SO FAR— AND OTHERS **WAITING** TO BE **EXPLORED.**

GROTTE DI FRASASSI, A KARST CAVE SYSTEM IN GENGA, ITALY

Bet You Didn't Know!

People often get confused by the different types of rock formations found in caves, but there is an easy way to remember the difference between a stalactite and a stalagmite. Stalactites (with a *c*) hang from the ceiling, while stalagmites (with a *g*) grow up from the ground!

BIG MACs IMPACT EARTH

No, we're not talking about hamburgers here! MACs refers to Meteoroids, Asteroids, and Comets, and all three have had a huge impact on our planet. Asteroids are large chunks of rock or metal that can range in size from hundreds of miles to a few dozen feet across. Most of the largest asteroids orbit the sun between Mars and Jupiter in a place called the asteroid belt, but there are those that come close to Earth and occasionally ram into us.

Comets can also orbit the sun, but instead of being made of rock and metal, they are made mostly of ice, small rocks, and dust with a smattering of other stuff mixed in. Astronomers believe that many comets initially come from way out in the solar system in a place called the Oort cloud. When comets get close to the sun, the frozen mass starts to melt, forming a large glowing cloud around the comet (called a coma) and a long tail made of glowing gas and dust that stretches out way behind it.

Meteoroids are small pieces of rock and metal. They can come from pieces broken off of asteroids, or they can be from the rocky bits found in a comet's tail. Some meteoroids can even be pieces of another planet or moon that got blasted off into space when it was hit by a MAC! Because they are relatively small, most meteoroids burn up as they fly through Earth's atmosphere, creating a streak of light that is called a meteor. If they survive the fall and are later found on Earth, they are then called meteorites.

When the solar system was young there were a lot more of these big MACs flying around and smashing into our planet. Asteroids brought some of the native iron found on Earth's surface, and some scientists think

COMET ISON AS SEEN THROUGH A TELESCOPE FROM EARTH

WHERE HAVE ALL THE **CRATERS GONE?**

Compared to our moon, Earth does not have many craters. The question is ... why?

Most meteoroids never make it to Earth's surface. Anything that's smaller than a car will usually burn up or explode in the air, creating a spectacular fireball. Those are the meteors you see at night that folks sometimes call shooting stars.

Only a few dozen objects that are the size of a pickup truck or larger strike the planet each year, and while they can leave a crater, many of them fall in places where people don't see them: Antarctica, large deserts, and the ocean, which covers almost three-quarters of Earth's surface. If craters do form, they usually fill in with dirt, and plants start to grow over them, so you don't really see that they are there. The reason that Meteor Crater in Arizona, U.S.A., on the opposite page looks so fresh is because it's in the desert, so there is very little water to wash it away and few plants can grow over it.

Geologists believe that Meteor Crater in Arizona formed about 50,000 years ago when an asteroid hit the planet. The crater is about 4,000 feet (1,200 m) across and 600 feet (180 m) deep.

that a lot of the water that we have on our planet was brought here by comets and asteroids (but this is still a topic of debate). What we do know for sure is that over the course of the past 4.5 billion years or so, Earth has been hit periodically by some pretty large objects that many scientists believe wiped out much of the life on the planet at the time. Yep, when it comes to dynamic processes that impact Earth, a large asteroid strike is about as big as it gets!

All meteorites are not created equally! Iron meteorites are made almost entirely of the metals iron and nickel. Stony meteorites, which is the largest group, have very little metal in them and look very similar to some Earth rocks. Then there are stony-iron meteorites, which—as the name suggests—have both iron and rock in them.

STONY-IRON METEORITE

CHAPTER THREE

THE **LIFE** AND **DEATH** OF **ROCKS**

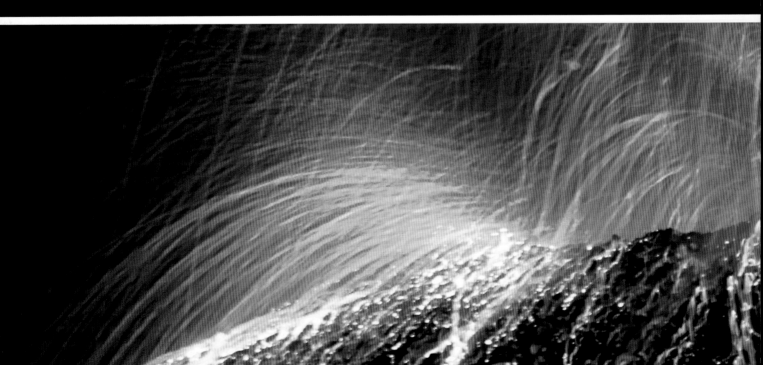

Even though rocks are not alive, they are "born" by different processes. And almost all of them eventually get destroyed and "die"—but can then be "born" again. The atoms and molecules that make up their minerals do live on, though, when old rocks get recycled into new rocks.

SOME LIKE IT HOT

Iceland's Svartifoss waterfall is surrounded by columns made of basalt, an extrusive igneous rock.

Do you think that rocks are hot stuff? Well, when some rocks first form, they are so hot they're not even solid. Igneous rocks begin their lives as a red-hot mass of molten material under the ground called magma. The word "igneous" comes from the Latin word *ignis*, which means "fire," and that's a perfect way to describe these rocks.

Igneous rocks come in two different forms, depending on whether they form above or below Earth's surface. Intrusive igneous rocks form inside Earth directly from pools of magma that cool slowly over time. When the magma is still a hot liquid, atoms and molecules in it can move around freely, but as the magma cools they lose energy and slow down. Eventually the atoms link up, forming chemical bonds, and the first mineral crystals start to grow. As the magma continues to cool, more minerals form, until the entire mass turns solid. Depending on its size, it can take thousands of years for a pool of magma to completely turn solid, and some never do. Because the cooling process is so slow, intrusive igneous rocks, like granite (pp. 78–79) and gabbro (pp. 82–83), have large mineral crystals in them.

Igneous rocks can also form on Earth's surface. These are called extrusive or volcanic rocks. As the name suggests, volcanic rocks come from volcanoes, which are formed when magma comes spewing out of a crack in Earth's crust. Once magma starts flowing over the surface, it is referred to as lava.

Bet You Didn't Know!

Magma rises to the surface because it is less dense than the surrounding solid rock. Most magma also has a lot of gas dissolved in it, which—just like the gas in a bottle of soda—causes it to erupt after it has been shaken and you take off the top. Magma erupts when the pressure of the gas in it becomes too great.

> **INTRUSIVE IGNEOUS ROCKS** ARE ALSO CALLED **PLUTONIC ROCKS—** NAMED FOR PLUTO, THE MYTHOLOGICAL ROMAN **GOD OF THE UNDERWORLD.**

MAKING **MAGMA**

So, what causes rocks to get so hot that they melt and make magma? Some of the heat comes from the radioactive decay of certain chemical elements that are concentrated in pockets in Earth's crust and mantle. Toward the end of the 19th century, several scientists including Marie Curie, her husband, Pierre Curie, and Henri Becquerel discovered that chemical elements such as uranium and thorium are very unstable and, over time, naturally change, or "decay," into other, more stable elements. When they decay, they release radioactivity, a type of energy that can generate a lot of heat—enough heat to even melt solid rock!

This same idea is used in nuclear power plants. Radioactive material is used to turn water into steam, which is used to power generators.

You can actually visit places to see extrusive igneous rocks forming today. One of the best spots is at Kilauea volcano in Hawaii, U.S.A., where rivers of molten lava flow right into the ocean and sometimes right over unlucky people's homes. Lava cools much faster than magma underground. As a result, extrusive igneous rocks like basalt (p. 90) have crystals that are so small you often can't see them without a microscope.

Whether they form deep underground or on Earth's surface, igneous rocks are indeed hot stuff!

ROCKY TRANSFORMERS

As you might expect, hot pools of magma rising through Earth and tectonic plates pushing into each other can really stress out the rocks in the crust. Sometimes these stresses cause rocks to break apart, but if the conditions are right, instead of breaking, the rocks completely change their form. Metamorphic rocks begin with igneous, sedimentary, or even other metamorphic rocks and then undergo a major transformation due to extreme heat and pressures that are usually found deep under Earth's surface.

Unlike igneous rocks, which start as molten rock, metamorphic rocks never totally melt. Instead, the changes happen in a solid state. Geologists have discovered that metamorphism can happen in a few different ways, which lead to different types of metamorphic rocks being formed. Contact metamorphism happens when magma flows up through Earth and the minerals in the "host rocks" surrounding the magma become altered. High temperatures "cook" the minerals, causing their bonds to break and their atoms to reorganize themselves, often forming totally new minerals. It's sort of like when you make toast in the toaster. The bread undergoes a change when it is heated but it never melts. Dynamic metamorphism, on the other hand, is caused mostly by directed pressure. This can either happen in a small area such as when rocks grind past each other along an active fault, or it can be

METAMORPHISM CAN HAPPEN ON EARTH'S SURFACE, BUT IT IS RARE. **IMPACT METAMORPHISM** HAPPENS WHEN EXTRA-TERRESTRIAL OBJECTS SUCH AS **ASTEROIDS** SMASH INTO EARTH.

Bet You Didn't Know!

Metamorphic rocks behave a little like plastic modeling clay. When the clay is cold it easily breaks apart when it is stretched, but if you roll it around in your hands for a while and get it warm, you can easily stretch it and bend it without it ever breaking.

Metamorphic rocks, like this gneiss, have undergone some big changes, getting squeezed and heated deep under the surface of Earth.

ALTERED STATES

So, just how much can a rock change when it undergoes metamorphism? Let's take a look! The first rock pictured below is called shale (p. 127). It formed underwater when layers of mud built up and then hardened into stone. The second is called garnet schist (p. 107), and it shows what happens when shale undergoes big-time metamorphism!

The transformation of shale into schist is called high-grade metamorphism, because there was a total change in the rock. The amount of change that a rock goes through is controlled by the type of metamorphic process that is acting on it, how long that process acts, and the type of rock that you are starting with. Rocks that undergo low-grade metamorphism look very similar to what they did when they started, but with higher heat and more intense pressure, the greater the change that happens in the rock.

large-scale regional metamorphism, which happens when rocks get caught between two tectonic plates that are ramming into each other. This type of metamorphism is clearly seen along most of the major mountain chains in the world, including the Alps in Europe and the Himalaya in Asia. When rocks undergo dynamic metamorphism, the pressure coming from a specific direction causes their minerals to become stretched and lined up in patterns called foliation.

Sometimes rocks can simply undergo metamorphism as a result of many layers of rock being deposited on top of them. This is called burial metamorphism, and it's the combination of pressure from above and natural heat trapped in the earth that causes the minerals in the host rock to change.

SHALE

SCHIST

TEARING DOWN AND BUILDING BACK UP

They may *look* like they will last forever, but even the tallest mountains and the hardest rocks will eventually wear away. Rocks at Earth's surface are constantly being broken down and weathered by wind, water, ice, and living things like lichens. The broken pieces of rocks are called sediment, and when pieces of sediment get joined back together again, they make a new sedimentary rock.

Rocks made from pieces of sediment are called clastic rocks, and they are the most common type of sedimentary rock. After it forms, sediment gets carried away in a process called erosion. Any moving fluid can cause erosion, including wind (yes, wind is a fluid in gas form!) and ice in glaciers, but most of the sediment that makes new sedimentary rocks gets transported by running water in streams and rivers. Smaller pieces of sediment, like sand, silt, and clay, are usually picked up and carried by the moving water directly.

Larger sediment chunks such as gravel and cobbles are too heavy to float in the water, but if the water is flowing fast enough, they can be pushed along the bottom where they roll and bounce along. That's why many of the rocks you see at the bottom of a stream are round. All the rolling and bouncing knocks off their sharp edges.

When rivers finally empty into lakes or the ocean, the sediment starts to pile up. This process is called deposition. Sometimes the sediment gets deposited in a jumbled mass, but most often it forms nice smooth layers. Dig a hole in the sand at the beach and you can see these layers. As the layers of

ABOUT **220 MILES** (355 KM) WIDE, THE **GANGES DELTA** HAS BEEN BUILT BY **SEDIMENT** CARRIED BY THE GANGES AND BRAHMAPUTRA RIVERS INTO THE BAY OF BENGAL ALONG THE COUNTRIES OF **INDIA** AND **BANGLADESH.**

Bet You Didn't Know!

Glaciers are dirt-moving machines! When a glacier flows, it acts like a conveyor belt, carrying in and on top of it everything from tiny grains of silt to boulders that are bigger than a car. The sediment that the glacier is carrying then gets dumped at the front of the glacier where it melts, forming a jumbled mass called a moraine.

Sediment eroded from the middle of North America is carried by the Mississippi River into the Gulf of Mexico, where it may eventually turn into new sedimentary rock deep under the seafloor.

AN UPLIFTING
EXPERIENCE

Even though most clastic sedimentary rocks form from the deposition of sediment underwater, you can often also see them way up high in the mountains. So how did they get there?

After the rocks formed, tectonic forces in Earth pushed them up, sometimes thousands of feet above sea level. One classic example of this can be found in the rocks of the Catskill Mountains in New York, U.S.A.

About 400 million years ago, this area was covered by a large inland sea, into which rivers carried sediment eroded from a large chain of mountains to the east. As the water flowed toward the sea, changes in sea level and the amount of water flowing in the streams caused the sediment to be deposited in thick layers, which eventually became sedimentary rock. Today the mountains that were originally in the east are completely gone, but their sediment lives on in the rocks of the reborn Catskill Mountains!

sediment build up, the grains at the bottom of the pile get squeezed and cemented together, creating a brand-new sedimentary rock!

There is a second type of sedimentary rock that doesn't form from pieces of rocks. They are called evaporites, and, as the name suggests, they form from the evaporation of water. When water runs over rocks, it can also dissolve some of the minerals. It's these dissolved minerals that make seawater taste salty. When salty water is trapped in pools in hot dry areas, or along the edge of the sea, the water evaporates, and new minerals begin to precipitate out. This is how minerals like halite (pp. 186–187) and gypsum (pp. 200–201) form.

UNREAL
LANDSCAPES

You might think that the processes of weathering and erosion just wear rocks flat, but like an artist with a hammer and chisel, they can carve out some incredible features—some of which look like they have come right out of a fantasy movie. Here are five of the most unreal landscapes that Mother Nature has sculpted out of rocks.

THE MATTERHORN, SWITZERLAND

Located in the Alps, the Matterhorn is one of the most famous mountain peaks in the world. It is also a great example of how glacial ice can carve even the hardest types of rock. Glaciers often form at the top of very tall mountains where snow piles up year after year, eventually turning into ice. When the ice gets thick enough, it slowly flows downhill, carving out chunks of rock as it moves. The Matterhorn gets its classic pyramid shape from the glaciers that over thousands of years carved out rounded bowls called cirques on each side of the mountain. They left a narrow ridge with a center spire called a horn sticking up.

THE WAVE AT VERMILION CLIFFS NATIONAL MONUMENT, ARIZONA, U.S.A.

It does not take much imagination to see why this incredible feature is called the Wave. If it weren't made of rock, you might just want to grab a surfboard and try to ride it! The wave is made of sandstone (p. 127) that was originally deposited by wind, making large dunes. The fantastic colors come from staining caused by iron and manganese oxides in the water that seeped through the sand grains after they were deposited, cementing them together. The Wave is believed to have been cut mostly by wind erosion; it demonstrates how nature's "sandblasting" can create some awesome landscapes.

SHILIN (STONE FOREST), CHINA

From a distance, the Shilin Stone Forest Geopark near Kunming, China, looks like a vast forest of pointed gray conifer trees. It's only when you get up close that you begin to realize that the "trees" are actually individual stone pillars, some of which rise up 150 feet (50 m) from the ground (the name Shilin means "stone forest" in Chinese). These spectacular structures formed from a thick layer of limestone that was deposited when the region was covered by a shallow sea some 270 million years ago. After tectonic forces lifted the area, flowing water and wind carved out the individual pillars, creating one of the world's best examples of eroded limestone. These features are so special that Shilin was named a UNESCO World Heritage Site in 2007.

HOODOOS AT BRYCE CANYON NATIONAL PARK, UTAH, U.S.A.

Each year millions of visitors who visit Bryce Canyon National Park in Utah are amazed by these odd-shaped towers of colorful limestone (pp. 120–121), sandstone (p. 127), and mudstone (p. 126) called hoodoos. The sediments that formed these rocks were originally deposited in a large lake about 50 million years ago and, after they hardened, were pushed up by tectonic forces. The carving of the hoodoos started when water seeped into cracks in the rocks, froze, and began to split them apart. Running water and wind did the rest!

THE FAIRY CHIMNEYS OF CAPPADOCIA, TURKEY

Some people think they look like giant mushrooms. Others say that they remind them of the spires of great cathedrals. But, most often, they are simply called fairy chimneys. Located in the Anatolia region of the country of Turkey, these giant towers of rock have been amazing travelers for thousands of years. The chimneys got their start millions of years ago, when volcanic eruptions blanketed the area with hundreds of feet of ash that eventually hardened to form a soft rock called tuff (p. 91). The tuff was then covered with a layer of basaltic (p. 90) lava, which formed a harder "caprock" on top. Over millions of years, fractures in the surface allowed water to flow down through the basalt, slowly eroding the underlying tuff. Later stream erosion wore away additional rock, creating today's "chimneys" that any fairy would be proud of!

FANTASTIC FOSSILS

FOSSIL OF AN
ANCIENT FERN

When many people hear the word "fossil," they think of the skeletons of huge dinosaurs that roamed Earth millions of years ago. And though there are plenty of awesome dino bones to go around, they are far from the only fossils found on our planet.

A fossil is the remains of any ancient life-form that has been preserved in rock. Animal fossils of all sizes—from tiny insects to giant whales—have been unearthed. Fossils of plants, fungi, microbes, and just about any other living thing that you can think of have also been found.

Most fossils are found in sedimentary rocks, which makes sense, because these remains would normally be destroyed by the intense processes that form both igneous and metamorphic rocks. Also, it's usually the hard parts of the organism that are preserved. Things like shells, teeth, wood, and bones stand up well when they get buried in sediment that eventually turns to stone. While it is possible to find fossils of soft parts of animals like skin and feathers, they tend to be rare.

Fossils come in several varieties. Some of the most common fossils are called molds and casts. Here's how they work: Bury a seashell in the mud and then wait for a few thousand years for the mud to turn to stone. Then allow water to flow through the rock and dissolve the shell away, leaving a space where the shell was. This space is the mold. The mold gets filled with new sediment, which turns to stone—making a perfect copy of the original shell. The copy is the cast.

Sometimes, a bone or piece of a tree trunk gets buried, and mineral-rich groundwater flows in and around it. Bone and wood have lots of little holes in them that the minerals can fill in and, over time, harden into stone. This process is called permineralization. Petrification is a similar process, except that when a piece of wood becomes petrified, the original organic material is replaced by new minerals.

DINOSAUR
TRACKS

Probably the coolest type of fossil happens when a tiny critter like an insect or tree frog gets caught in a big blob of sticky resin from a tree. Over time, the resin will harden, making a rock called amber (pp. 244–245) with the entire creature perfectly preserved inside!

Bet You Didn't Know!

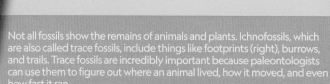

Not all fossils show the remains of animals and plants. Ichnofossils, which are also called trace fossils, include things like footprints (right), burrows, and trails. Trace fossils are incredibly important because paleontologists can use them to figure out where an animal lived, how it moved, and even how fast it ran.

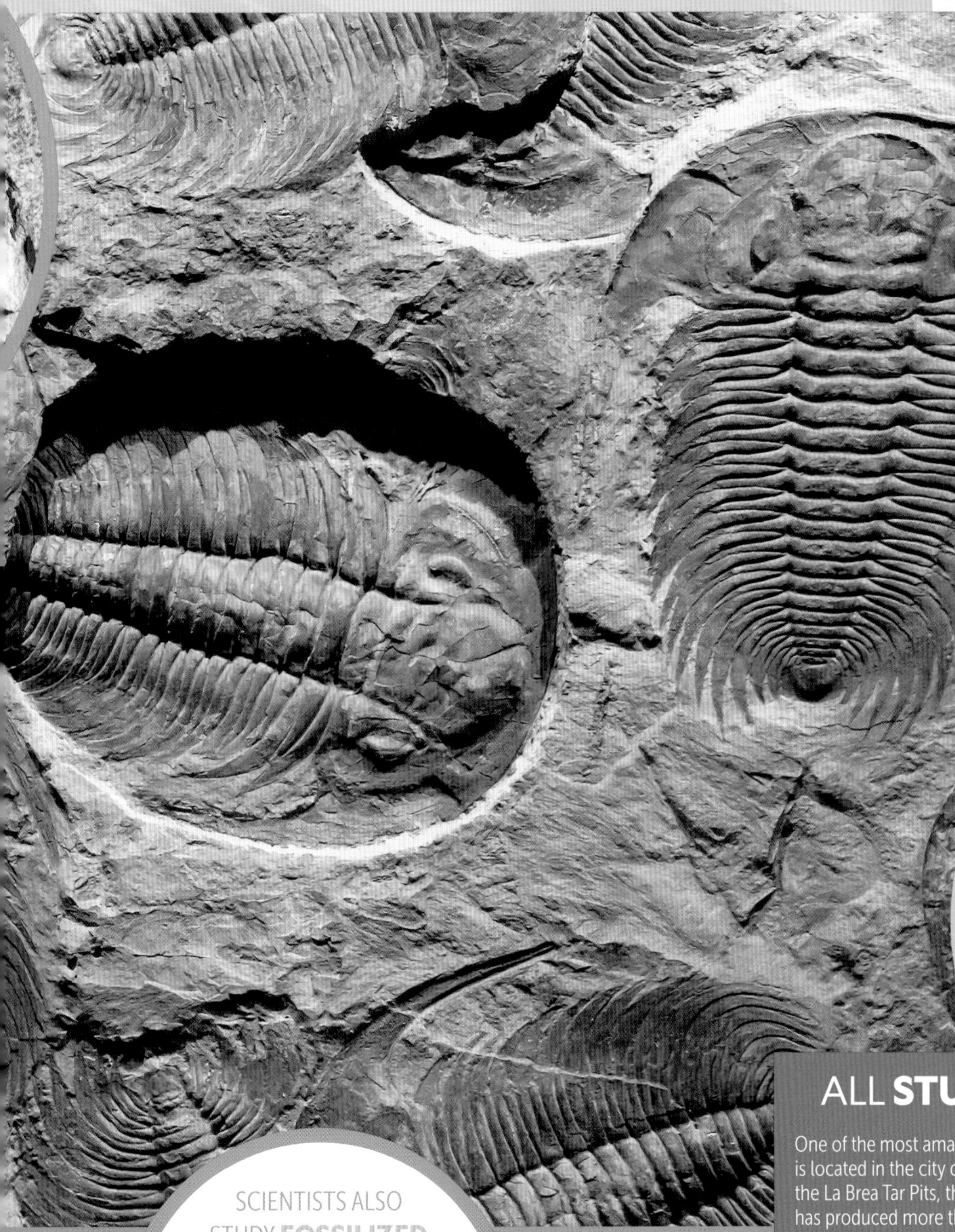

These imprints in sediment were made by trilobites, now extinct marine animals that existed from about 542 to 251 million years ago.

SCIENTISTS ALSO STUDY **FOSSILIZED POOP,** KNOWN AS **COPROLITES.** IT MAY SOUND **GROSS,** BUT PALEONTOLOGISTS CAN USE COPROLITES TO FIND OUT MORE ABOUT THE **DIETS** OF **EXTINCT ANIMALS.**

ALL **STUCK UP!**

One of the most amazing places to find fossils is located in the city of Los Angeles. Known as the La Brea Tar Pits, this treasure trove of fossils has produced more than 3.5 million specimens of Ice Age animals and plants, with more being found every day!

Rancho La Brea is a place where asphalt—a supersticky form of petroleum—naturally seeps out of the ground, forming pools on the surface. During the end of the last ice age, the area was home to all sorts of amazing animals including dire wolves, mammoths, mastodons, and saber-toothed cats. Water would cover the asphalt pools, and animals that came to take a drink—thinking it was a pond—would suddenly find themselves stuck. Predators looking for an easy meal would also get stuck and, over the centuries, more and more animals met an untimely death. The city of Los Angeles was built up around it. The La Brea Tar Pits now feature a park and a museum that people can visit.

FORWARD INTO THE PAST

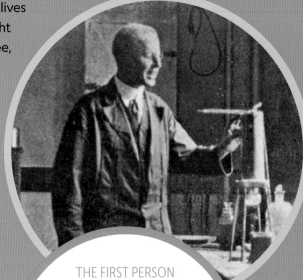

Age is a relative thing. Compared with a mouse that only lives a few years, an African elephant that is 60 years old might seem really old. But compared with a giant redwood tree, which can live to be 2,000 years old, a 60-year-old elephant is really young!

Before the early 20th century, most people believed that Earth was less than 10,000 years old. Then, in 1905, English physicist Ernest Rutherford suggested that a process called radiometric age dating could be used to get an accurate age for some types of rocks. Using it, scientists calculated that Earth is closer to 4.54 billion years old. (See the sidebar for an explanation of how radiometric age dating works.)

Even as far back as the 1700s, some scientists were not convinced that Earth could be only a few thousand years old. One was James Hutton, who came up with an important geologic principle that we now call uniformitarianism. Uniformitarianism is usually explained by the phrase "the present is the key to the past," which means the processes that are happening on Earth today can be a guide to how things changed in the past. By carefully observing the way rocks changed, Hutton realized that a bunch of small changes acting over a really long time could lead to some really big changes in the Earth!

Uniformitarianism is put to good use figuring out how old different layers of the Earth are. Layers of the Earth are called strata, and stratigraphy looks at how different rock layers stack up and how their ages compare with one another. Some of the rules of stratigraphy are straightforward. The first says that if you have a bunch of rock layers that have not been disturbed, then the layer at the bottom is the oldest. Another says that if you have one type of rock cutting across another rock, then it's the rock doing the cutting across that's younger.

THE FIRST PERSON TO SUCCESSFULLY USE **RADIOMETRIC AGE DATING** TECHNIQUES WAS AMERICAN SCIENTIST **BERTRAM BOLTWOOD.** IN 1907 HE DATED A SET OF ROCKS AT **2.2 BILLION YEARS OLD.**

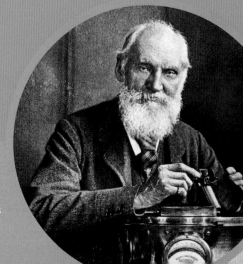

Bet You Didn't Know!

One of the first people to calculate Earth's age using a scientific approach was British physicist William Thomson, more commonly known as Lord Kelvin. In 1863, he published a paper explaining that if Earth had started as a molten mass, then based on the current temperature of Earth's rocks the planet would have to be at least 100 million years old. His number was totally wrong, but in many of his calculations, he showed that Earth was much older than most people had believed.

GEOLOGIC TIMESCALE

ERA	PERIOD		EVENTS
CENOZOIC	Quaternary		Evolution of humans
	Tertiary		Mammals diversify
MESOZOIC	Cretaceous		Extinction of dinosaurs First primates First flowering plants
	Jurassic		First birds Dinosaurs diversify
	Triassic		First mammals First dinosaurs
PALEOZOIC	Permian		Major extinctions Reptiles diversify
	Carboniferous	Pennsylvanian Mississippian	First reptiles Scale trees Seed ferns
	Devonian		First amphibians Jawed fishes diversify
	Silurian		First vascular land plants
	Ordovician		Sudden diversification of Metazoan families
	Cambrian		First fishes First chordates
LATE PROTEROZOIC			First skeletal elements First soft-bodied metazoans First animal traces

MILLIONS OF YEARS AGO: 0, 18, 50, 100, 150, 200, 250, 300, 350, 400, 450, 500, 550, 600, 650

One of the things that really helped with linking up rocks of the same age was the use of index fossils. These special fossils are from living things that lived for only a short period of time but can be found in sedimentary rocks all over Earth. For example, say you find a fossil of a trilobite in rocks that you know formed between 520 and 510 million years ago. And then, on the other side of the world, your friend digs out a fossil of the same species of trilobite. Because your friend knows the age of the trilobite, they can deduce that the rocks they found it in are between 520 and 510 million years old, too. Using the rules of stratigraphy and index fossils to show that different rocks were the same age, geologists were able to construct a single geologic timescale that covers the age of the entire planet.

CLOCKS IN THE ROCKS

Radioactive decay provides the heat to melt rock into magma and fuel nuclear reactors, but it can also be used to figure out how old igneous rocks are. When radioactive elements naturally break down, or decay, they do so at a set rate of time called the half-life. Here's how it works.

When a mineral first forms from cooling magma, it has a fixed amount of certain radioactive elements in it. These elements are known as the parents. Parent elements will begin to decay at a constant rate and change into different elements known as the daughter elements. The half-life is the amount of time needed for half of the parent element to change into a specific daughter element. By knowing how long the half-life is, and by comparing the amount of parent elements with daughter elements found in a rock, you can calculate how long ago the mineral formed and how old the rock is. Today this process is called radiometric age dating.

WHAT **GOES** AROUND **COMES** AROUND

I f you measure the ages of all the rocks on Earth, you'll find that there are only a few that date back to the earliest days of our planet. You may be wondering why this is. The answer is simple. It's called the rock cycle. Because of the different dynamic processes at work on and inside Earth, most of the rocks that formed in the past have been recycled into new rocks. In some cases this has happened dozens of different times. Check out how the different parts of the rock cycle play out.

GRANITE

MAGMA

1

Heat inside Earth causes some of the rocks to melt, forming magma. Sometimes this magma hardens underground, forming new intrusive igneous rocks.

BECAUSE THERE IS **NO ACTIVE ROCK CYCLE** OPERATING ON **EARTH'S MOON,** EVEN THE **YOUNGEST ROCKS** FOUND ON ITS SURFACE ARE **BILLIONS OF YEARS OLD.**

UPLIFT **2**

Tectonic forces at work deep inside Earth raise the land surface, exposing deeply buried rocks to weathering and erosion.

WEATHERING, EROSION, AND DEPOSITION **3**

Flowing water, moving glaciers, and wind break up rock to make sediment, which is then transported and deposited in low-lying areas such as lakes and oceans.

EVAPORATION **4**

Pools of mineral-rich water slowly evaporate, leaving behind crystals of new sedimentary rock.

CONGLOMERATE

LITHIFICATION **5**

Buried sediments get compacted and cemented together, forming new sedimentary rocks.

VOLCANIC ACTION **6**

Lava flowing out of erupting volcanoes hardens to form new extrusive igneous rocks.

Subduction zone

TECTONIC ACTION **7**

The motion of Earth's tectonic plates forces rocks together where they get heated and squeezed, creating new metamorphic rocks.

GNEISS

GEARING UP

OK, fellow rock hounds:

Now that you have the basic information you'll need to go out on your own geologic adventures, you will need to gear up so that you'll be able to identify and collect the rocks and minerals that you encounter in the field. Here's a list of some of the tools of the trade that we geologists use to get the job done right!

SAFETY TIP!

Always check with a trusted adult before you head off on a geologic adventure!

STRONG MAGNET
Some rocks and minerals, like magnetite (pp. 172–173), contain iron, so you want to test them to see if they are magnetic.

HAND LENS
A simple pocket magnifier will allow you to get a closer look at the crystals or grains in a rock.

MEDICINE DROPPER BOTTLE FILLED WITH VINEGAR
Some rocks like limestone (pp. 120–121) and minerals like calcite (pp. 190–191) react and bubble when vinegar is placed on them.

ROCK HAMMER AND CHISEL
These are used for breaking off a small sample to take home with you and for exposing a fresh surface of the rock to see the minerals better.

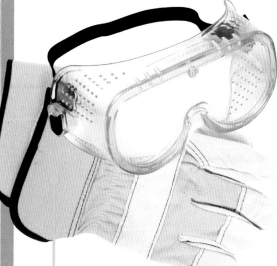

GLOVES AND EYE PROTECTION
Sturdy leather gloves and either goggles or safety glasses protect you while you're hammering away.

STEEL NAIL FILE
This is a handy tool for testing the hardness of minerals and for sharpening your pencil when the point breaks.

NOTEBOOK AND PENCIL
It's always a good idea to take plenty of notes when you are in the field. Information about location, date of visit, and even weather conditions may come in handy later on.

CELL PHONE
You should always go collecting with a responsible adult, but you never know when you may need to call home. Most smartphones also have a camera for taking pictures of the landscape and a GPS built in, just in case your responsible adult gets you lost!

STRONG BACKPACK
You're going to need something to carry all your gear plus the cool samples that you collect!

CAMERA
Sometimes a rock is too large to take home, or you want to remember the setting that it was found in. You know what they say: A picture is worth a thousand words!

WATER BOTTLE
This helps keep you hydrated when you're working in the field, plus the water comes in handy if you want to clean off a sample with your paintbrush.

SMALL PORCELAIN TILE
The unglazed back of a porcelain tile will allow you to test the streak of a mineral when you rub the two together.

PAINTBRUSH
This will allow you to clean off some of the dirt covering a rock to get a better look.

71

CHAPTER FOUR
SMOKIN' HOT
IGNEOUS ROCKS

When volcanoes like Mount Pinatubo in the Philippines erupt, they are spectacular sights to behold. But the material they eject can often bring big changes to Earth's surface.

GETTING THE SCOOP ON IGNEOUS ROCKS

There are several keys that geologists use to identify igneous rocks. Because all igneous rocks start with a melted mass of magma, they are all made of minerals that have grown together and have an interlocking structure. This immediately separates them from sedimentary rocks, most of which are made from individual particles or grains that have been cemented or squeezed together. Also, when mineral crystals are found in igneous rocks, they are usually randomly distributed throughout the rock. Metamorphic rocks contain many of the same minerals as igneous rocks, but the crystals are often lined up in patterns. The problem is, extrusive igneous rocks cool so fast that the crystals are either microscopic or nonexistent, so you can't always see them.

Igneous rocks can have different groups of minerals in them. It turns out that not all magma is created equally. Each type of magma has its own unique chemical composition, but they tend to fall into two main groups. Felsic magmas are rich in the elements silicon, aluminum, and oxygen but have far less of the elements iron, calcium, and magnesium in them. As a result, felsic rocks are generally light in color and contain the minerals quartz (pp. 206–207) and feldspars (pp. 212–213) with very little olivine (pp. 234–235) and

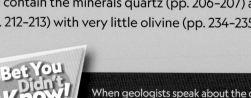

GABBRO

PERIDOTITE IS AN ULTRAMAFIC IGNEOUS ROCK THAT IS THOUGHT TO BE A **MAIN COMPONENT** OF **EARTH'S MANTLE.**

GRANITE

Bet You Didn't Know!

When geologists speak about the grain size of an igneous rock, what they are describing is the size of the crystals that make up the rock. Sometimes this can be confusing because, technically speaking, grains are pieces of sediment like sand or silt and are most often found in sedimentary rocks.

Even though all igneous rocks form from the cooling of hot molten rock, they can look very different from one another depending on where they formed.

BASALT

LET'S TALK ABOUT **TEXTURE**

Geologists use many different terms to describe the texture, or size of the crystals, in igneous rocks. Rocks with really big crystals are called pegmatites, and they form from magmas that flow very easily, allowing minerals to grow fast. Rocks with crystals that are clearly visible to the naked eye such as granite (pp. 78–79) and gabbro (pp. 82–83) have a phaneritic texture. These rocks form from magmas that cool slowly so the crystals have time to grow. Rocks that require some type of magnification to see the crystals have an aphanitic texture. These rocks cool very quickly. They include basalt (p. 90) and rhyolite (p. 90). Rocks like a phonolite porphyry (pp. 76–77) that have a mixed texture with both large crystals and tiny crystals together have a porphyritic texture.

Volcanic rocks can also have some unusual textures. Scoria (p. 91) and pumice (pp. 88–89) have no crystals, and since they are full of holes, they have a vesicular texture. When obsidian (pp. 86–87) forms, it cools so fast that the minerals do not have time to form crystals. It is said to have a glassy texture. When chunks of volcanic material get welded together, the rocks have a pyroclastic texture.

pyroxenes (pp. 224–225). The most common felsic rocks are in the granite (pp. 78–79) family. These are mostly found making up the continental crust.

Mafic magmas are pretty much the exact opposite of felsic magmas. They contain lots of heavy elements such as iron and magnesium and far less silicon and aluminum. As a result, mafic rocks tend to be dark colored and have lots of the minerals pyroxene and olivine but very little quartz. The most common mafic rocks are in the basalt (p. 90) family and are found mostly in oceanic crust, although they do show up on the continents, too.

As you might expect, there are some magmas that fall between felsic and mafic composition. These "intermediate rocks" are most often found along the edges of continents near active and extinct volcanoes and include diorite (pp. 80–81) and andesite (p. 91). And then there are rocks that have mostly heavy elements. These ultramafic rocks are so dense they are believed to have formed in Earth's mantle and have been carried up into the crust by magma flowing up toward the surface.

OBSIDIAN

THE DEVIL IS IN THE DETAILS

Does this rock formation look familiar? You may have seen its iconic shape in movies or read about it in books and magazines. A favorite of photographers all over the world, Devils Tower is located in the Black Hills region of the U.S. state of Wyoming. Rising up almost 600 feet (200 m) from the grassy plain below, the tower looks like an oval-shaped mountain that is almost flat on top. Geologists have been studying the tower for more than 100 years and they are still not certain how it formed. Here's what we know for sure.

The main rock making up the tower is a relatively rare type of igneous rock called a phonolite porphyry. It is in between mafic and felsic in its composition and is composed of very small crystals of pyroxene (pp. 224–225) minerals and larger crystals of white orthoclase feldspar (p. 213), with very little quartz (pp. 206–207). The rocks of Devils Tower are about 50 million years old, having cooled from a molten mass of magma that pushed its way up into layers of sedimentary rocks that once covered the area. Originally, the rocks of the tower were formed deep under the surface. But millions of years of erosion have stripped away most of the sedimentary rocks that were on top of it and in the area surrounding it. The reason that the tower is still standing is because the igneous rock that it's made from is very resistant to the forces of weathering and erosion. The big question that geologists have is what the rocks of the tower looked like when they first formed. Some say the

ALL CRACKED UP

Some of the most striking features of Devils Tower are the long vertical columns that can be seen peeling off the sides of the tower itself. When you get close to them, you can see that many of these columns are massive, with some being almost 10 feet (3 m) across and hundreds of feet high. But what's truly unusual about the columns is their shape. Some are hexagonal, meaning they have six sides, while others are five-sided. Still others are irregularly shaped. They all formed by a process called columnar jointing, creating a geologic structure found only in igneous rocks.

Since the rock making up Devils Tower formed deep under the surface, it cooled very slowly. As it cooled, it also contracted—meaning that it got smaller. This caused it to crack in a honeycomb-like pattern. Once the tower was exposed at the surface by erosion, the cracks started to split apart, causing the columns to peel away and pile up at the bottom.

Admired by millions of people each year, Devils Tower is made from igneous rock that originally formed deep underground.

tower resembled an igneous intrusion geologists call a stock, which is shaped like a large rectangular prism. Others say that it was a mass called a laccolith, which is more mushroom shaped, and a few geologists think it might be the remains of a mass of magma that plugged up the neck of a volcano. Unfortunately, we'll probably never know the exact story because the rocks that had the information needed to complete the picture have long since disappeared!

ANOTHER COOL ROCK FORMATION THAT SHOWS THE SAME TYPE OF **COLUMNAR JOINTING** AS DEVILS TOWER IS **GIANT'S CAUSEWAY,** LOCATED IN NORTHERN IRELAND.

Bet You Didn't Know!

In 1906, then president Theodore Roosevelt recognized the importance of the rocks in and around Devils Tower by declaring the site the first of America's national monuments, guaranteeing that its unique geology would be protected for future generations.

GRANITE

Granite makes up the foundation of much of the continental crust in the world, so it is an extremely common rock. It is an intrusive igneous rock, and its minerals crystallize as magma that is rich in the compound silica cools deep under Earth's surface. Because the cooling takes place slowly, the mineral crystals are clearly visible to the naked eye. Sometimes, granites will have exceptionally large crystals. These rocks are called granite pegmatite.

The main minerals found in granite are potassium and plagioclase feldspars (pp. 212–213), with at least 10 percent quartz (pp. 206–207). In addition, both muscovite (p. 217) and biotite (p. 217) are common, as well as the amphibole horn-blende (p. 228). The mineral makeup controls the color of the rock, which can be quite variable. When the rock contains a lot of potassium feldspar such as orthoclase, granite looks pink or even red, but when plagioclase feldspars are the dominant type, the rock looks gray or has a classic salt-and-pepper appearance, which is white with lots of black specks.

Granite can be found in many mountain areas where it has been pushed up to the surface by tectonic forces and erosion has removed the softer rock around it, making it stand out. It can also be seen in veins cutting across other rocks, especially when it is in the form of pegmatite. Because it is a hard rock, granite often forms spectacular peaks such as in Torres del Paine National Park in Chile.

The main use of granite is in construction and, when it is polished, as a decorative stone. It is also used to make tombstones and monuments, dating back thousands of years to the Egyptian pyramids. Granite can also often be found as paving stones in older roads. Because of both its attractive crystals and its durability, people often use polished granite as countertops in their homes.

FACTS

ROCK TYPE: igneous—intrusive, felsic composition

MAIN COLORS: gray, white and black, pink, red

TEXTURE: mostly phaneritic with moderately large crystals .08–0.2 inch (2–5 mm)

STRUCTURE: generally none but may show slight layering of minerals

HARDNESS: greater than 5.5

COMMON MINERALS: potassium feldspar, quartz, plagioclase feldspar, muscovite, biotite, hornblende

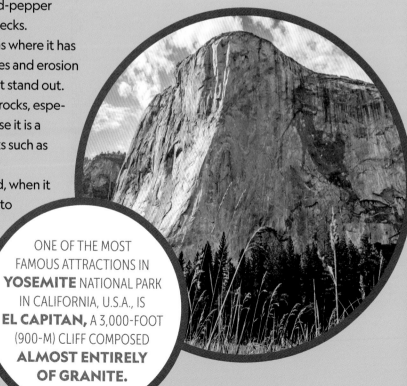

ONE OF THE MOST FAMOUS ATTRACTIONS IN **YOSEMITE** NATIONAL PARK IN CALIFORNIA, U.S.A., IS **EL CAPITAN,** A 3,000-FOOT (900-M) CLIFF COMPOSED **ALMOST ENTIRELY OF GRANITE.**

Mineral crystals are clearly visible in granite, which comes in a variety of colors.

NATURAL BRIDGE GRANITE ROCK FORMATION CAUSED BY COASTAL WAVE EROSION IN TORNDIRRUP NATIONAL PARK, WESTERN AUSTRALIA

Bet You Didn't Know!

Many of the rocks sold as granite for use in countertops and other decorative stone are not really made of granite. They are frequently other coarse-textured igneous rocks. To be a true granite, a rock must contain at least 10 percent quartz; rocks sold as black granite have almost no quartz in them. Most often, they are diorite (pp. 80–81) or gabbro (pp. 82–83).

Diorite is a darker colored igneous rock that can have a salt-and-pepper appearance because of the large mineral crystals found in it.

Bet You Didn't Know!

Many statues, vases, and bowls made from polished diorite have been found in the tombs of ancient Egyptian rulers. But most of the cutting tools back then were made from copper or bronze, which are much softer than diorite ... so how were these items carved? One theory is that ancient Egyptians used other hard stones to chip out the basic design and then used simple hand drills tipped with a harder mineral such as corundum (pp. 166–167) to cut in the fine details.

DIORITE

Though not as common as granite (pp. 78–79), diorite can be found in many locations around the world. Like granite, it is an intrusive igneous rock, and its minerals slowly crystallize deep under Earth's surface from a magma that has a great deal of iron and magnesium in it. The mineral crystals are clearly visible to the naked eye, often featuring much larger crystals called phenocrysts made of either feldspar (pp. 212–213) or hornblende (p. 228) surrounded by smaller mineral crystals.

The main minerals found in diorite are plagioclase feldspars such as andesine, hornblende, and pyroxenes such as augite (p. 224). Diorite is usually made up of less than 5 percent quartz, making it denser than granite. It can also have small amounts of biotite (p. 217). Most often, diorite is dark greenish gray or has a salt-and-pepper appearance.

Diorite can be found along the edges of continents, frequently forming a rim around the edge of a large mass of granite that has been raised to the surface by tectonic forces. It can also form smaller layers called dikes that cut across other rock types. Because it is a hard rock resistant to weathering and erosion, diorite masses will often form raised ridges or domes on the land surface that tend to be rounded near the top.

The main use of diorite over the years has been as an ornamental stone to make monuments and sculptures. Because of its dark color and similar crystal structure, diorite is sometimes referred to as black granite and, after it is polished, is used for countertops or for stone that is applied to the face of buildings. Diorite is also sometimes mined and crushed to be used as an underlying material in roads or along curb cuts.

FACTS

ROCK TYPE: igneous—intrusive, intermediate between felsic and mafic composition

MAIN COLORS: dark greenish gray to black, white with black

TEXTURE: mostly phaneritic with moderately large crystals .08–0.2 inch (2–5 mm)

STRUCTURE: generally none but may show slight layering or banding of minerals

HARDNESS: 5.5–6

COMMON MINERALS: plagioclase feldspar, potassium feldspar, hornblende, pyroxene

BEFORE PEOPLE HAD **METAL CUTTING TOOLS,** THEY OFTEN USED **POLISHED DIORITE** FOR MAKING THINGS LIKE **CHISELS AND AXES.**

GABBRO

Gabbro is only rarely found in continental rocks, but it is a major component of the rocks that make up the crust of the ocean. Gabbro is an intrusive igneous rock formed from mafic magma that is rich in iron and magnesium. It usually forms below layers of basalt (p. 90) in the ocean crust, where it cools slowly enough for large crystals to grow. Most often gabbro appears to be dark gray or greenish black, but sometimes the crystals will be layered with alternating dark- and light-colored bands. The layering happens because some minerals in the magma crystallize at a higher temperature than the rest. When they turn solid, they sink to the bottom of the magma, forming a layer. Then the rest of the minerals crystallize, filling in the rest of the rock.

The main minerals found in gabbro are plagioclase feldspars (pp. 212–213) such as labradorite, pyroxenes such as augite (p. 224), hornblende (p. 228), and olivine (pp. 234–235). Depending on the exact chemical composition of the magma that it formed from, gabbro can also contain the minerals biotite (p. 217), magnetite (pp. 172–173), ilmenite (p. 175), and corundum (pp. 166–167). When present, quartz (pp. 206–207) is found only in tiny amounts.

Gabbro is not a common rock on the land surface, but it does sometimes occur as igneous intrusions cutting across other rocks, making saucer-shaped masses called lopoliths. Most often, it is found near the edge of a continent where parts of the ocean crust have been pushed up on the land surface.

Gabbro is cut and polished and used as a decorative stone on the outside of buildings and, like diorite (pp. 80–81), is sometimes called black granite. In other places, it is mined and crushed to be used as "trap rock" filler along railroad tracks and roadways. Sometimes, a gabbro will have a high concentration of magnetite (pp. 172–173) or ilmenite (p. 175), and it will be mined as an ore (a rock that contains a certain type of useful metal) of iron or titanium.

FACTS

ROCK TYPE: igneous—intrusive, mafic composition

MAIN COLORS: dark gray to black, sometimes greenish black

TEXTURE: phaneritic with moderately large crystals .08–0.2 inch (2–5 mm)

STRUCTURE: generally none, but layered gabbro is fairly common

HARDNESS: 5.5+

COMMON MINERALS: plagioclase feldspar, hornblende, pyroxenes, olivine

> TAKEN TOGETHER, **GABBRO** AND ITS EXTRUSIVE COUNTERPART, **BASALT** (P. 90), MAKE UP ABOUT **75 PERCENT** OF ALL THE **IGNEOUS ROCKS** IN THE **EARTH'S CRUST.**

Bet You Didn't Know!

Gabbro doesn't just form on Earth! In 2007, a large chunk of rock containing pieces of gabbro was discovered in the country of Morocco. What made this rock unique was that it was covered in something called fusion crust, meaning that it was actually a meteorite. After much investigation, scientists determined that this rock came from the highland area of the moon!

Gabbro is a dark-colored intrusive igneous rock that sometimes has a layered appearance, with large crystals that can easily be seen with the naked eye.

GABBRO ON THE ISLE OF SKYE, SCOTLAND

LOTS OF LAVA

When steaming lava comes pouring out of a volcano, it can make a bunch of different rock formations on the surface. Lava is hot stuff, often ranging in temperature from 1300 to 2200°F (700–1250°C)! The Hawaiian Islands exist because, over time, piles of lava have built up on top of the ocean floor, eventually breaking the sea surface. These islands offer visitors the opportunity to see some truly fantastic features formed by flowing lava.

KILAUEA VOLCANO IN HAWAII, U.S.A., **ERUPTED** ALMOST **CONTINUOUSLY** FOR **35 YEARS.**

PĀHOEHOE LAVA

One common type of Hawaiian lava is called pāhoehoe (pronounced "pa-hoy-hoy"), which looks like long strands of braided rope. Pāhoehoe is usually found near the top end of lava flows, forming large lobes or tongues that are fed by highly fluid lava from underneath. When the lava hits the air, the top surface cools, forming a thin skin that starts to bend and flop over on itself as new lava is added from underneath. Pāhoehoe can have a glassy surface, but because gas flows out of it as it cools, it also can have lots of little holes in it.

'A'Ā LAVA

The other major type of lava to form in Hawaii is called 'a'ā (pronounced "ahh-ahh"). The surface of 'a'ā lava is extremely rough and hard to walk on, often with big angular or blocky chunks sticking up and lots of burnt fragments called clinkers. The lava that forms 'a'ā is much thicker and less fluid than the lava that forms pāhoehoe, though a flow that starts out as a pāhoehoe flow near the top end can turn into an 'a'ā flow near the bottom. Instead of moving like a river, a flow of 'a'ā lava moves more like the treads of a tank with the top of the flow rising up and over and then falling down in front of the lava that has already cooled.

LAVA TUBES

Some lava flows are so hot and fluid that they can form long tunnels called lava tubes. As the lava flows, the top and sides cool and harden, forming a solid shell through which the liquid lava continues to flow and push the front of the flow forward. As the eruption dies down, the amount of liquid lava flowing in the tunnel is reduced. Eventually, just a long hollow tube is left. Some lava tubes, like the Thurston Lava Tube on the Big Island of Hawaii, can be quite spectacular, running for hundreds of feet and spanning more than 20 feet (6 m) in height.

PILLOW LAVA

It's easy to see why these rock formations are called pillow lavas. Their rounded shapes and big lobes look almost like someone is stocking up for a major league pillow fight ... though you wouldn't want to get hit by these pillows! Pillow lavas form from either slow-flowing lava that oozes up out of the ground or lava that erupts under the sea. In both cases, the pillows form because the outside of the lava cools quickly, forming a shell that slowly expands, almost like a balloon that is being inflated. These features are commonly found on the edge of the ocean.

PELE'S HAIR AND TEARS

Some eruptions of lava can be quite spectacular, producing actual fountains of lava that shoot high into the air. At the very top of these fountains, drops of red-hot liquid break off and freeze solid in the air, forming little round particles—called Pele's tears—that rain down onto the ground. (Yes, whenever a liquid—even lava!—turns solid due to cooling, it "freezes.") Sometimes, as the drops are turning solid, they get stretched out as they move through the air, forming long fibers that look like hair but are actually supersharp strands of glass. These long fibers are called Pele's hair. Both the tears (below) and the hair (left) are named for Pele, the Hawaiian goddess of volcanoes, fire, and lightning.

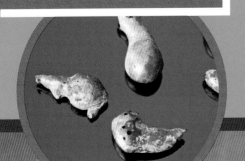

GO WITH **THE FLOW**

The basaltic lava found in the Hawaiian Islands is extremely fluid, so when it erupts, it flows quite easily over the ground surface and sometimes can travel for miles. This has caused some serious problems for people living in Hawaii. In the past, lava flows have blocked roads, cutting off entire communities, and have destroyed buildings. Fortunately, most lava flows are very predictable, and the folks at the Hawaiian Volcano Observatory monitor the eruptions and post warnings so that people can evacuate their homes if need be. Still, most people probably don't want to hear that a red-hot river of molten rock is heading for their front door!

OBSIDIAN CLIFF IN YELLOWSTONE NATIONAL PARK

Bet You Didn't Know!

The edge of a scalpel that has a chip of obsidian embedded in it can be sharper than one made from steel, so doctors will sometimes use them for surgery. It turns out that this is not a new technology. Some archaeologists believe that obsidian knives were used in ancient Egypt to prepare bodies that were being turned into mummies.

OBSIDIAN

When most people hear the word "glass," they usually picture something you drink from or a window that you look through. Humans have been making glass from silica sand for almost 6,000 years, but obsidian—nature's own glass—has been forming for billions of years. Just like manufactured glass, obsidian breaks with an extremely sharp edge. Because of this, people of the past used it to make cutting tools and weapons long before they started using metals.

Obsidian forms from felsic magma that is rich in silica and has very little water in it, so when it erupts from a volcano as lava, it is super thick and flows very slowly. As a result, the lava cools so quickly that no mineral crystals can even form. Technically speaking, the material that it's made from is called a mineraloid. In some cases, obsidian samples can be found with tiny quartz crystals called phenocrysts that formed before it cooled. You can also sometimes see lines in the glass made by other materials caught in the moving lava as it flowed.

Obsidian is commonly found near lava domes and lava flows that contain rhyolite (p. 90). Sometimes, it forms a rim around an igneous intrusion where cooling has occurred very quickly. One of the best locations for finding obsidian in the United States is at Obsidian Cliff in Yellowstone National Park, but it has been discovered all over the world, including in the Zemplén hills of Slovakia. Most obsidian is black, but it can also be reddish brown when there is iron present.

Besides serving as ancient tools and weapons, obsidian was also used by people in ancient Greece as well as by the Aztec in Mexico to make mirrors. Today, it is often cut and polished to make jewelry and other pieces of artwork. Yes, obsidian is a natural glass with a whole lot of class!

FACTS

ROCK TYPE: igneous—extrusive, usually felsic composition

MAIN COLORS: black, brown, reddish brown

TEXTURE: glassy; no visible crystals except as phenocrysts

STRUCTURE: some flow banding, shell-like fracture

HARDNESS: 5–5.5, but will scratch window glass

COMMON MINERALS: feldspar and quartz phenocrysts are possible

SNOWFLAKE OBSIDIAN

SNOWFLAKE OBSIDIAN IS BLACK OBSIDIAN WITH LOTS OF LITTLE **WHITE PATCHES** IN IT THAT LOOK ALMOST LIKE—YOU GUESSED IT—**SNOWFLAKES.**

PUMICE

Have you ever thrown a rock into a pond? Chances are it sank like a stone, because most rocks are much heavier than an equal amount of water. Pumice is the exception to that rule.

Pumice is the only rock that naturally floats because it is literally full of holes, many of which still contain gas in them. So the rock bobs around on the surface, almost like a piece of Styrofoam.

Pumice is an extrusive igneous rock that forms from felsic magma that is rich in silica. When the magma erupts as lava, it is often full of gas. As the gas is suddenly released, it forms bubbles, creating a froth on top of the lava flow, almost like the foaming head on a glass of root beer. This type of lava cools especially quickly and, when it does, the bubbles freeze (become solid) right in the rock. In some cases, some tiny minerals of feldspar (pp. 212–213), pyroxene (pp. 224–225), and hornblende (p. 228) can be seen with the help of a magnifier. Over time, some of the holes can even fill in with other minerals that have been deposited by water flowing through the rock when it is buried.

Pumice can come in a variety of colors, including white, yellow, brown, and black. The most common types of pumice form near volcanoes that have rhyolite (p. 90) lava flows, but they can also form near volcanoes that produce basalt (p. 90).

Over the years, people have found a variety of uses for pumice, including making concrete blocks that are both lightweight and strong. Pumice stones are used in landscaping and are also often ground up and used as an abrasive in cleaners, including some brands of hand soap. If you wanted to, you could even carve a piece of pumice to make a model boat, all thanks to the rock that floats!

FACTS

ROCK TYPE: igneous—extrusive, usually felsic composition

MAIN COLORS: black, white, yellow, brown

TEXTURE: glassy; visible crystals rare

STRUCTURE: lots of small holes or vesicles

HARDNESS: 6

COMMON MINERALS: feldspars, hornblende, pyroxene

PUMICE CAN BE USED TO HELP REMOVE **DEAD SKIN** FROM ELBOWS, KNEES, AND THE BOTTOM OF FEET. IT'S LIKE **NATURAL SANDPAPER** FOR THE SKIN!

Bet You Didn't Know!

One of the largest volcanic eruptions to happen in modern times was in 1883, when the Krakatoa volcano—located in Indonesia on an island between Java and Sumatra—blew its top. Not only was it one of the deadliest eruptions, but many Indonesian harbors became jammed with floating chunks of pumice. Pumice doesn't just come from volcanoes on land, though. In 2012, a huge raft made of floating chunks of pumice was discovered in the Pacific Ocean. The source? A submarine volcano near New Zealand called the Havre Seamount!

Pumice is light enough to float on the surface of this glass of water. The heavier geode (pp. 208–209) sinks to the bottom.

EXTRUSIVE EXCLUSIVE!

When magma cools slowly deep underground, the minerals have lots of time to grow, and in most cases, produce large crystals. When lava erupts out of a volcano, the cooling happens very quickly, so even though many of these extrusive igneous rocks have the same chemical composition as their intrusive cousins, they look very different because they have much smaller crystals. Under some conditions, erupting lava will also produce some unusual textures and structures in a rock that cannot be formed underground. Here's an exclusive look at some of these "groundbreaking" extrusive igneous rocks!

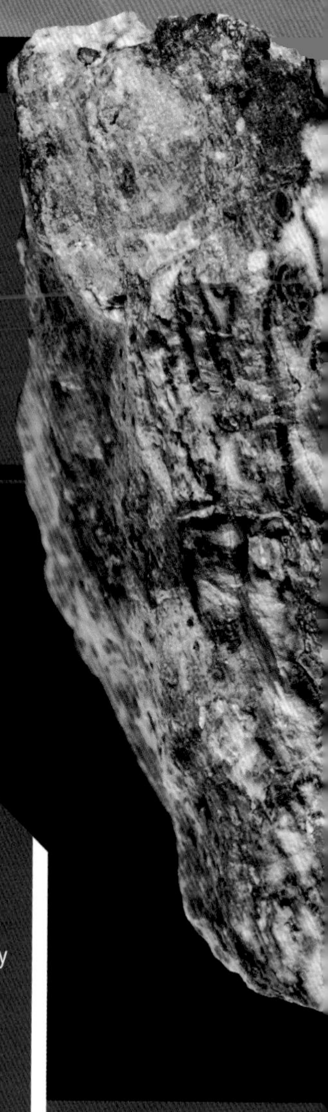

RHYOLITE

Rhyolite is a felsic rock that has the same chemical composition as its intrusive igneous cousin granite (pp. 78–79). And, like granite, it is commonly found on continents. Since it forms from quickly cooled lava, rhyolite is found near many volcanoes that often produce pumice (pp. 88–89) and obsidian (pp. 86–87). When it does form mineral crystals, they are usually very small, although in some cases there are large quartz crystals (pp. 206–207). Quite often rhyolite will show bands that come from the flowing lava that created it.

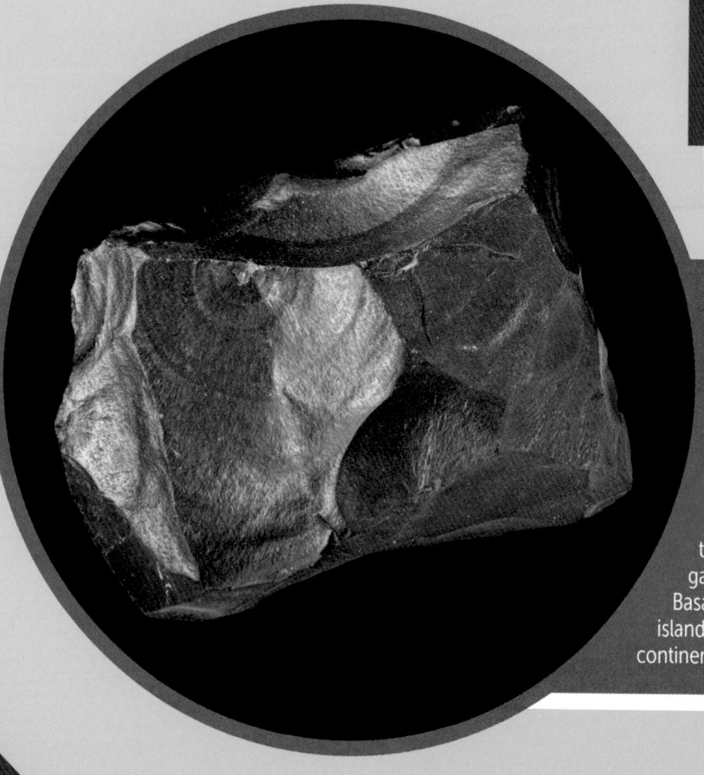

BASALT

Making up most of the oceanic crust on the planet, basalt is the most common rock found on the surface of Earth. Basalt is a dark-colored mafic rock, usually black, dark gray, or even greenish in color. It mostly has an aphanitic, or fine-grained, texture often with small mineral crystals of plagioclase feldspar (pp. 212–213), pyroxene (pp. 224–225), and olivine (pp. 234–235); and it is the extrusive equal of the intrusive igneous rock gabbro (pp. 82–83). Both gabbro and basalt have the same chemical composition. Basalt can be found making up many of the volcanic islands in the Pacific Ocean and at a few locations on the continents, where it has been deposited in massive lava flows.

TUFF

Most rocks are tough stuff and are hard to break, but volcanic tuff is not one of them. Tuff is a relatively soft rock that is easy to break because it's made from layers of volcanic ash and other solid ejecta that have become fused together over time. The texture and chemical composition of tuff can take many different forms, and it often has a layered structure.

SOME **VOLCANOES ERUPT** WITH **SUCH FORCE** THAT THEY HURL LARGE CHUNKS OF SOLID ROCKS CALLED **VOLCANIC BOMBS** HUNDREDS OF FEET INTO THE AIR.

ANDESITE

Even though andesite is an extrusive igneous rock, it can often be found with a porphyritic texture, or with both large and small crystals mixed together. In andesite, large mineral crystals of plagioclase feldspar (pp. 212–213) are often embedded in a mass of very small pyroxene (pp. 224–225) and/or biotite (p. 217) crystals. The rock gets its name from the Andes Mountains in South America where it is commonly found. It is also found in volcanoes that make up the Ring of Fire. Andesite has the same chemical composition as the intrusive rock diorite (pp. 80–81).

SCORIA

Some lava flows have a great deal of gas trapped in them as they flow out of a volcano. If the lava cools quickly enough, it produces a rough rock with lots of holes in it called a scoria. Scoria are dark colored and usually form from the same type of magma that makes basalt and andesite. Unlike pumice (pp. 88–89), which also has a lot of holes, it is too dense to float on water.

ROCK-SOLID HEADS OF STATE

Can you think of any other statue that looks like Mount Rushmore? The size and scale of the four presidential heads is staggering. Carved out of solid granite (pp. 78–79), the heads are each about 60 feet (18 m) high and, combined, stretch almost 200 feet (65 m) across the face of the mountain.

Even before he started carving the rock in October 1927, sculptor Gutzon Borglum knew that he wanted his mammoth tribute to U.S. presidents George Washington, Abraham Lincoln, Thomas Jefferson, and Theodore Roosevelt to last, so the rock type that he used was critical. After touring the area of the Black Hills of South Dakota, he selected Mount Rushmore and he knew that it was a good choice.

Much of the rock that makes up the core of the Black Hills formed more than 1.6 billion years ago, when a pool of felsic magma deep underground pushed its way up into even older sedimentary rock and cooled to form the granite from which the faces are carved. As it intruded, the magma cooked the surrounding rocks, transforming them into the metamorphic schist (p. 107) and quartzite (pp. 102–103) that surround the area today.

Geologists aren't quite sure what happened in the area over the next 1.5 billion years or so, because rocks from this time are no longer around. But they do know that, about 70 million years ago, tectonic forces inside Earth began lifting the region. This uplift caused most of the rocks covering and surrounding the granite to erode, or wear away, producing the mountain that we see today.

THE **NOSE** ON **GEORGE WASHINGTON'S** FACE AT **MOUNT RUSHMORE** IS **21 FEET** (6.4 M) LONG, AND EACH **EYE** IS **11 FEET** (3.4 M) WIDE.

Because granite is so hard, carving the faces required a great deal of dangerous work. First, workers would hang from chairs suspended by cables and place dynamite into holes that they drilled into the rock. They then blasted layer after layer of rock off the face until they got to within about six inches (15 cm) of the final surface. Then the workers used jackhammers to take off the rest. Final smoothing was done with something called a bumper tool. It took about 400 workers almost 14 years to finish the monument, which will hopefully last thousands of years into the future!

GETTING A **FACE-LIFT**

Even though they are carved from super-resistant granite, the great stone faces of Mount Rushmore still face wear and tear from the forces of weathering and erosion. Geologists estimate that the rate of erosion is only about 1/10 inch (2 mm) for every thousand years, but a major concern comes from frost wedging. As it cooled from magma, the granite contracted, creating tiny cracks in the rock. When water seeps into the cracks and then freezes, it expands, splitting the cracks open.

To help reduce this problem, a team of workers regularly climbs the faces to inspect the rocks for new cracks, which they then fill with a special type of silicon caulk to keep water from getting in. Since 1998, they have been helped by a high-tech system of wires and fiber-optic cables that acts as an early warning system, allowing them to monitor the rocks for new cracks.

Bet You Didn't Know! Even though it is not open to the public, there is a large chamber called the Hall of Records cut into the rock behind Abraham Lincoln's head at Mount Rushmore. In its floor, covered by a granite capstone, is a box that contains copies of the Declaration of Independence and the U.S. Constitution that have been carved into porcelain tiles.

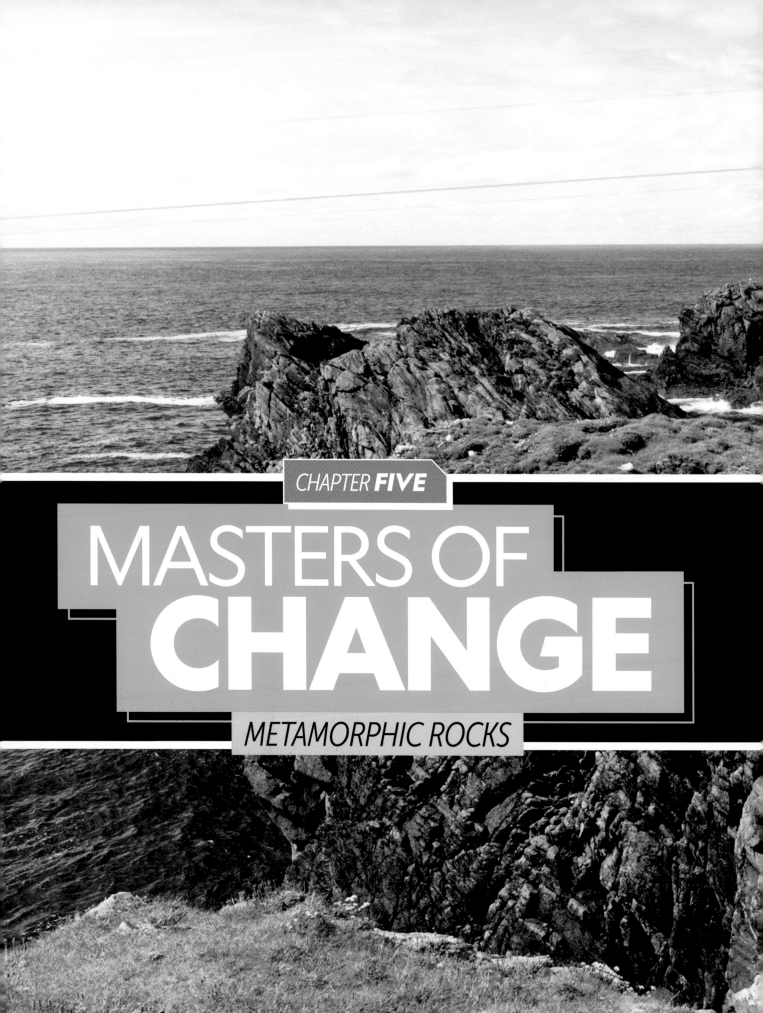

CHAPTER FIVE

MASTERS OF CHANGE

METAMORPHIC ROCKS

Metamorphic rocks come in several different varieties, but the minerals and structures found in each of them tell the story of how the rocks have changed over time.

THE SQUEEZE ON METAMORPHIC ROCKS

MARBLE

The term "metamorphic" comes from the Greek terms *meta,* meaning "change," and *morph,* meaning "form." And that's exactly what metamorphic rocks have done: They've changed their form as the geologic conditions around them have also changed.

All metamorphic rocks start out as either an igneous, a sedimentary, or a different metamorphic "parent" rock. When the conditions surrounding the parent rock undergo a big change (like when it's caught between colliding tectonic plates or gets hit by an asteroid), the rock changes, too. Over time the minerals in the parent rock recrystallize, and, as they do, they preserve a record of both the original rock and the different processes that have affected it.

Like igneous rocks, metamorphic rocks are made up of crystals that have fused together. The more time these minerals have to grow, the larger the crystals in the rock tend to be. The big difference between metamorphic and igneous rocks is that, unlike igneous rocks, metamorphic rocks don't totally melt. Instead, they recrystallize while in a solid state when they are heated to a high temperature or when hot mineral-rich fluids or magma flows through them. Quite often, metamorphic rocks that are closer to a molten mass of magma have a different set of minerals in them than those that are farther away, even if they start from the same parent rock. Because some minerals form only at certain temperatures, geologists can use these

When it comes to metamorphic rocks, the original parent rock will often control the mineral makeup. Pure limestone (pp. 120–121), a sedimentary rock that is made from only calcite (pp. 190–191), will produce a marble. But a "dirty limestone" that also has some clay and sand in it will often produce a calc-silicate rock with a totally different set of minerals in it, including garnets (pp. 242–243) and diopside (p. 224).

Metamorphic rocks like this garnet schist show how rocks can change when they are heated and squeezed under enormous pressures found inside Earth.

"index minerals" to get clues about the conditions that were present when a rock underwent metamorphism.

Because many metamorphic rocks also have been under a great deal of pressure, the minerals frequently will line up in patterns that geologists call foliation. If a rock has no mineral alignment, then either the pressure was spread out evenly (like when a rock gets buried deep underground) or all the change was due to changes in temperature. If the pressure comes from a specific direction, the mineral pattern can tell geologists which direction the pressure came from. A gneiss (pronounced "nice") (p. 106), for example, will have bands of different minerals running through it, and a migmatite (p. 107) will display distinct folds showing that the pressure came from one side or the other. A schist (p. 107) or phyllite (p. 106), on the other hand, will have minerals like muscovite (p. 217) and biotite (p. 217) layered in sheets.

NEVER TAKE A **GNEISS** ROCK FOR GRANITE!

Both igneous and metamorphic rocks are made from interlocking crystals, and they often contain the same minerals, so they can sometimes be hard to tell apart. That's where the structure of the rock comes into play. Take granite (pp. 78–79), for example. It is a felsic rock that has a great deal of quartz (pp. 206–207), biotite (p. 217), and both potassium and plagioclase feldspar (pp. 212–213). It turns out that gneiss may have the exact same group of minerals in it. So how can you tell the two apart? Because gneiss formed under pressure, the minerals in it form distinct bands that can sometimes even be folded over and bent. In granite, the mineral grains are randomly spread throughout the rock. While this difference may seem small, it is an important clue that will keep you from taking a gneiss rock for granite!

GEOLOGISTS CALL THE ORIGINAL **PARENT ROCK** OF A METAMORPHIC ROCK THE **PROTOLITH.**

GRANITE

GNEISS

WHITE GOLD FROM ITALY

When most people hear the word "gold," the first thing they think of is the yellow metal used to make expensive jewelry. But the "gold" we're talking about here is a white rock known as Carrara marble. It comes from the Apennine Mountains of northern Italy, and to some craftspeople, it's as valuable as real gold. Marble is a metamorphic rock that has long been used by builders and artists because it is both durable and easy to cut. Most forms of marble have impurities in them and are streaked with lines of different colors, but pure marble—like the stuff found in the Massa-Carrara province of Italy—can be snow-white, making it extremely valuable for use in sculptures. Some of Italy's greatest artists, including Michelangelo and Leonardo da Vinci, sought out Carrara marble when they were creating their works, and it was used in the 1850s by Giovanni Strazza when he carved his magnificent statue called "The Veiled Virgin" (left).

The marble quarries found in Massa-Carrara are easy to spot. From a distance, they look like big patches of snow or ice covering some of the mountaintops. But when you get up close, you can see that they are actually giant pits that have been cut into the hillsides exposing the marble below. Quarry workers use heavy machinery to peel away giant chunks of marble from the sidewalls of the excavation site, and special saws to cut them into huge blocks, some of which can weigh more than 20 tons (18.1 t)—more than a large school bus that is fully loaded with kids! The blocks are then stacked for storage or transported to workshops, where they are then cut into smaller slabs

The Italian sculptor Michelangelo used Carrara marble for some of his greatest works, including his statue "David" (right) and the "Pietà." Legend has it that he would often travel to the quarry himself to pick out the exact stones that he would use in his sculptures.

Snow-white marble quarried in Massa-Carrara province in Italy has been used for sculptures and buildings for thousands of years.

A **TOWERING** ACHIEVEMENT

The quarries in the Apennine Mountains have supplied builders with decorative stone for more than 2,000 years. The marble has been used in some of the most spectacular structures ever built, including Trajan's Column.

Built entirely out of marble for the Roman emperor Trajan between the years A.D. 106 and 113, this magnificent tower stands 126 feet (38 m) high and is covered with intricate carvings detailing the history of the wars that he had fought earlier in his reign. What makes the tower unique is that it was built from almost two dozen enormous cylinders of Carrara marble stacked one on top of the other. Each cylinder was cut from a single block of marble and had a section of a staircase carved directly into the center of the rock. When the cylinders were put in place, it formed one continuous spiral staircase on the inside that went all the way to the top!

to be turned into floors, countertops, staircases, and, of course, sculptures.

Unfortunately, like many natural resources, Carrara marble does not have an endless supply. Many of the quarries in the region have already been abandoned because they have run out of the best quality stone. As the stone becomes rarer, it is possible that sometime in the future pure Carrara marble will be truly worth its weight in gold!

THE **IMAGES CARVED** ON THE OUTSIDE OF **TRAJAN'S COLUMN** READ LIKE **ONE CONTINUOUS COMIC STRIP** THAT WRAPS AROUND THE COLUMN. FROM BEGINNING TO END, IT IS MORE THAN 800 FEET (240 M) LONG!

MARBLE

From the Washington Monument in Washington, D.C., to the Taj Mahal in India, when architects and builders were looking to make their stone structures really stand out, their rock of choice was quite often gleaming white marble.

Marble has a few properties that make it ideal for use as a decorative stone. First, it is a very common rock type, with large deposits found throughout the world. In many cases, the rock found in marble deposits is very consistent, which means that when rocks are removed from the quarry, they look very similar to one another. This is a big plus when you are covering a large building with stone and you want the individual pieces to match. Because marble is a soft rock compared to something like granite (pp. 78–79), it is easy to work with and can be split, cut, and carved into different shapes. Finally, when it is polished, marble will really shine, especially when the sun hits it just right!

Marble forms when either limestone (pp. 120–121) or dolostone (pp. 194–195) are heated and squeezed under a lot of pressure, usually as a result of being buried deep under thick layers of other sedimentary rocks. These conditions cause the original calcium-rich sediments to recrystallize, forming a massive rock that has a granular texture of interlocking calcite (pp. 190–191) or dolomite (pp. 194–195) crystals. Marble can also form by contact metamorphism, when a large mass of magma flows through or around the limestone, cooking the rock and causing the new mineral crystals to form. During metamorphism, almost all the original sedimentary features found in the parent rock (including layering and fossils) get destroyed.

Sometimes, the parent limestone will not be pure, but instead will have thin layers of impurities such as clay, silt, or sand mixed in with the limey material. In these cases, the marbles will often have colorful streaks and other structures in them and may also contain small amounts of other minerals including diopside (p. 224), tremolite (p. 229), garnet (pp. 242–243), and sometimes even quartz (pp. 206–207).

FACTS

ROCK TYPE: metamorphic—usually contact or regional metamorphism

MAIN COLORS: gray, white, pink

TEXTURE: granular with crystals up to .08 inch (2 mm)

STRUCTURE: usually none but may show some streaking due to impurities

HARDNESS: 3

COMMON MINERALS: calcite, dolomite

THE **LINCOLN MEMORIAL** IN WASHINGTON, D.C., IS MADE FROM **DIFFERENT TYPES OF MARBLE** THAT COME FROM THE U.S. STATES OF ALABAMA, GEORGIA, AND COLORADO.

Most marble tends to be gray, white, or even pink, and it can quite often have colorful streaks due to materials like silt and clay found in the original limestone that it formed from.

Statues made from pure white marble almost look like they are lit up from the inside. When the marble's surface is polished, light from the outside penetrates the outer surface of the stone and bounces off the faces of the calcite crystals below.

Bet You Didn't Know!

Quartzite is made almost completely from the mineral quartz.

Bet You Didn't Know!

Sandstones are made from grains of sand that have joined together and have lots of tiny spaces in them called pores. Sometimes, water flowing through the pores will fill them with minerals, making the rocks super heavy. Sometimes these types of rocks are called sedimentary quartzites, and the metamorphic type are called metaquartzites, but most often the term "quartzite" is used for the metamorphic kind.

QUARTZITE

Quartzite is a very hard but brittle rock. It is made almost entirely of interlocking quartz (pp. 206–207) crystals that form when a sandstone (p. 127)—which is made up almost entirely of quartz sand—undergoes either contact or regional metamorphism in which hot, mineral-rich fluids flow through the rock. Most often a rock needs to have at least 90 percent of its minerals be quartz for it to be called a quartzite, but sometimes this figure is closer to 99 percent. If the original parent rock had small amounts of silt and clay mixed with the sand, the quartzite might have small amounts of garnet (pp. 242–243) and muscovite (p. 217) in it, too.

The easiest way to tell a quartzite from a quartz sandstone is to look at the grains. Under heat and pressure, the original sand grains in the sandstone recrystallize and lose their rounded shape along with any layering that might have been present in the parent rock. When you hit a piece of quartzite with a hammer it splits right through the crystals, leaving a smooth surface, often with a sharp edge. When you hit a sandstone, it breaks around the grains, often leaving a rough edge. The way quartzite breaks made it a very useful resource for people of the past who turned the rock into cutting tools such as hand axes or scrapers. In some cases, it was also used to make hammerstones.

The tight crystalline structure of quartzite makes it very resistant to weathering and erosion. It often forms ridges in areas where other metamorphic rocks such as marble (pp. 100–101) and schist (p. 107) have been worn away. Large quartzite outcrops and cliffs are common, with lots of pointed crags sticking up. Today, quartzite is mostly used in the construction of roads and railroad beds. It is sometimes crushed and used to make specialized silica brick. When it is cut and polished, quartzite is also used as decorative stone for things like steps and countertops.

FACTS

ROCK TYPE: metamorphic—either regional or contact metamorphism

MAIN COLORS: white, light to dark gray, occasionally pink

TEXTURE: fine to medium-size crystals 1/16–3/16 inch (2–5 mm)

STRUCTURE: usually found in uniform masses with no foliation

HARDNESS: 7, but is brittle and breaks with an uneven fracture

COMMON MINERALS: over 90 percent quartz

WARNING!

YOU MUST ALWAYS WEAR **GLOVES AND EYE PROTECTION** WHEN YOU HIT ROCKS WITH A HAMMER BECAUSE **CHIPS** CAN GO **FLYING EVERYWHERE.** THIS IS ESPECIALLY TRUE FOR **QUARTZITE,** WHICH IS **REALLY BRITTLE!**

AMPHIBOLITE

As the name suggests, amphibolite is a rock that is chock-full of amphibole minerals, which are dark in color because they are rich in the elements iron and magnesium. Most typically, these rocks will contain large percentages of hornblende (p. 228), which is usually black, and either tremolite (p. 229) or actinolite (p. 229), which tend to give the rock a greenish color. Unlike quartzite (pp. 102–103), which is made almost entirely of pure quartz, an amphibolite can also contain a variety of other minerals including plagioclase feldspars (pp. 212–213), pyroxenes (pp. 224–225), calcite (pp. 190–191), and even some garnets (pp. 242-243). Amphibolite has a variable, or inconsistent, structure. Sometimes it appears as a large mass with crystals randomly scattered or with a small amount of alignment. When rocks that are rich in amphiboles are strongly foliated with the minerals forming distinct lines, it is usually called a hornblende schist.

Amphibolite is one of the most common rocks found in places that have experienced both high temperatures and pressure over a large geographic area. In many cases, the rocks are thought to have been formed deep underground at the bottom of mountains that were later pushed up by tectonic forces. Amphibolite can form from two very different types of parent rock. Most commonly, the parent is a mafic igneous rock such as a gabbro (pp. 82–83) or basalt (p. 90), but they can also form from sedimentary rocks such as "dirty" limestones (pp. 120-121) and graywackes, which are sandstones that contain large amounts of silt and clay.

Amphibolite feels heavy compared to other rocks that are usually found on the continents, such as granite (pp. 78–79) and gneiss (p. 106), because it is chock-full of heavy chemical elements like iron and magnesium. This property makes it ideal for use as filler in the construction of both highways and railroad beds. Amphibolite is also used as a decorative stone when it is cut and polished. Sometimes it is used on the outside of buildings, but, because it is relatively hard, it also makes great floor tiles. On occasion, amphibolite has been mined as a source of garnet (pp. 242–243), but these are not used as gems. They are used to make sandpaper and other abrasive materials.

FACTS

ROCK TYPE: metamorphic—medium- to high-grade metamorphism

MAIN COLORS: gray to black, sometimes green

TEXTURE: fine to medium-size crystals 1/16–3/16 inch (2–5 mm)

STRUCTURE: often massive with random crystal orientation but can be highly foliated

HARDNESS: 5–6

COMMON MINERALS: amphiboles, plagioclase feldspar, pyroxenes, calcite, rarely quartz

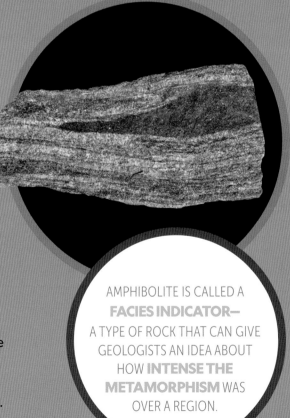

AMPHIBOLITE IS CALLED A **FACIES INDICATOR—** A TYPE OF ROCK THAT CAN GIVE GEOLOGISTS AN IDEA ABOUT HOW **INTENSE THE METAMORPHISM** WAS OVER A REGION.

Amphibolite is a heavy, dark-colored metamorphic rock with crystals that are clearly visible to the naked eye.

Bet You Didn't Know!

Because they form in areas where tectonic plates crash together to form mountains, amphibolites are often found with other types of metamorphic rocks that have undergone this same type of regional metamorphism. It is very common to find large bands of amphibolite next to bands of quartzite (pp. 102–103), marble (pp. 100–101), and gneiss (p. 106) in the same area.

PRESSURE COOKER

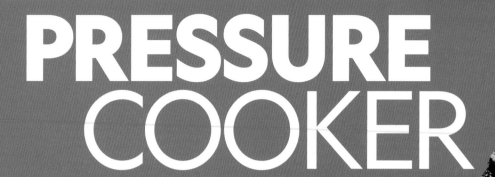

Sometimes, when parent rocks undergo metamorphism, the change doesn't happen all at once. As heat and pressure increase, they will undergo a slow, steady change from one type of metamorphic rock to another. This chain of events often happens with a sedimentary rock that formed from layers of silt and clay. Let's see what happens when we start with a typical shale (p. 127) as the parent rock goes through metamorphism at higher and higher temperatures and pressures.

1 SLATE

Slate has a dull finish and comes in a variety of colors including black, blue, green, and even purple. It is usually the first type of metamorphic rock to form when a fine-grained sedimentary rock such as shale (p. 127) is buried and heated to a low temperature of up to about 400°F (200°C). As the original shale is slowly cooked, its minerals change into tiny crystals of mostly mica minerals such as muscovite (p. 217), biotite (p. 217), chlorite (p. 217), quartz (pp. 206–207), and possibly some feldspar (pp. 212–213). Because of the pressure, the minerals line up in planes in the rock, producing zones of weakness—areas along which the rock can easily be split into thin sheets. This is known as slaty cleavage, and it makes slate perfect for use as roofing tiles and in patios.

2 PHYLLITE

To the untrained eye, slate and phyllite look very much alike. Both form from fine-grained sedimentary rocks, and both display well-developed cleavage that causes them to break into sheets. The main difference between the two is that a phyllite has been heated and squeezed for a longer period of time, causing the mica crystals in it to grow large enough to be seen with the naked eye. This gives the surface of phyllite a silvery sheen that slate does not have. Also, when phyllite splits apart it does not always form nice thin sheets. Instead, the broken pieces often form thicker slabs that have a jagged edge and an irregular shape.

3

SCHIST

When you turn up the heat a little and really increase the pressure on a clay-rich sedimentary rock, the next level of metamorphic rock that forms is schist. In a schist, the mineral crystals are quite large and show a distinctive layering, called schistosity, that causes the rock to split into irregularly shaped chunks. Schists usually have large flakes of different types of mica (pp. 216–217) along with some talc (pp. 220–221) with only small amounts of quartz (pp. 206–207) and feldspar (pp. 212–213). There are different types of schist featuring different minerals depending on the original chemical composition. Some of these include muscovite schist, garnet schist, talc schist, and biotite schist.

4

GNEISS

When clay-rich sedimentary rocks are heated to really high temperatures and squeezed under high pressures, the product is a metamorphic rock called a gneiss (pronounced "nice"). The thing that sets a gneiss apart from a slate, phyllite, or schist is that it does not have a tendency to split into sheets and is, in fact, hard to break. The different minerals form light and dark bands with different-size crystals in each. When it does break, it is in a random fashion. The main minerals found in a gneiss are quartz (pp. 206–207) and feldspar (pp. 212–213), and it may also include biotite (p. 217), hornblende (p. 228), and sometimes even garnets (pp. 242–243). Gneiss often is found in the cores of ancient mountain chains that have been eroded away, and they can also have granite (pp. 78–79) as their parent rock.

5

MIGMATITE

So, what happens when a rock becomes so hot it begins to partially melt again? You get a migmatite, a mixed-up rock that's part igneous and part metamorphic and that has lots of cool patterns running through it. These hybrid rocks often form in places where felsic magma has intruded or flowed into the parent rock. Many migmatites display the same type of mineral bands of quartz (pp. 206–207), feldspar (pp. 212–213), and biotite (p. 217) as a gneiss, but there will be veins of granite running through the rock, too.

Bet You Didn't Know!

When metamorphic rocks form, the heat and pressure that they go through usually destroy most of the structures and features that were part of the parent rock. Hornfels is an exception. Sometimes, when hornfels forms from a sedimentary rock that has alternating layers of sandstone (p. 127) and siltstone, the hornfels will end up having bands of minerals that line up with those original layers found in the parent rock.

HORNFELS

While many rocks have spectacular crystals, intricate patterns, or fantastic colors, hornfels is not one of them. It has a dull luster, no visible mineral crystals, and usually comes in either dark gray or black. Despite its less than exciting looks, however, hornfels is an important rock type because of what it can tell geologists.

Hornfels is one of the few types of metamorphic rocks where pressure has no role in its change, so when the minerals recrystallize, they are not aligned or foliated in any way. Hornfels forms strictly by contact metamorphism as a result of being very close to a large amount of magma that flowed into the parent rock, so it gets baked at a very high temperature, often as high as 1470°F (800°C). Because of the environment it forms in, hornfels often contains minerals such as andalusite (p. 241) and cordierite, which only form at very high temperatures. The problem is that none of the minerals are visible to the naked eye (though some samples do have large garnet [pp. 242–243] crystals in them). Most of the time, to see individual crystals, you need some type of serious magnification. The outer surface of the rock is usually smooth and quite solid. Individual rocks feel heavy because the minerals contain many heavy elements and they are tough to break, even with a hammer and chisel. When they do break, the pieces are usually sharp and angular.

Hornfels can form from many different parent rocks, including sedimentary shale (p. 127), igneous basalt (p. 90), and even other metamorphic rocks such as a slate (p. 107) and schist (p. 107). Unlike regional metamorphic rocks that can cover very large areas, hornfels are usually found very close to the mass of magma that heated them and caused them to form. This helps geologists get a handle on just how hot the magma was. Yes, even rocks that seem boring at first can prove to be very interesting in the long run!

FACTS

ROCK TYPE: metamorphic—high-grade contact metamorphism

MAIN COLORS: dark gray to black

TEXTURE: microscopic crystals

STRUCTURE: massive with random crystal orientation but can have leftover banding due to sedimentary layering of the parent rock

HARDNESS: 6–7

COMMON MINERALS: andalusite, cordierite, hornblende, plagioclase feldspar, sometimes garnet

HORNFELS, LIKE AMPHIBOLITE, IS CALLED **A FACIES INDICATOR** BECAUSE IT IS USED BY GEOLOGISTS TO TELL **HOW HOT** THE **PARENT ROCKS** GOT WHEN **MAGMA** FLOWED NEAR THEM.

SOAPSTONE

Most rocks are much harder than metals such as copper, brass, and even steel, but there is one type of rock that people can carve with a simple butter knife! It's called soapstone, and it gets its name from the fact that it feels kind of slippery, just like a bar of soap. It is one of the softest rocks there is. In fact, soapstone is so soft that you can scratch it with your fingernail!

Soapstone is super soft because its main mineral component is talc (pp. 220–221), the softest mineral on the planet. Soapstone usually forms when a dark, dense igneous rock called peridotite undergoes low temperature and low pressure metamorphism. In addition to talc, soapstone may also include pyroxenes (pp. 224–225), micas (pp. 216–217), chlorite (p. 217), and possibly some amphiboles (pp. 228–229). Soapstone commonly is found along the edge of continents where tectonic action has uplifted igneous rocks from deep in the crust and then heated them to the point where the original minerals have recrystallized.

More than 3,000 years ago, people discovered that they could use soapstone to make a variety of useful objects, including cups, cooking pots, wood-burning stoves, and bowls. Today, soapstone countertops are very popular in homes, although, unlike countertops made from granite (pp. 78–79), they can easily scratch. Soapstone is also used to create molds for making objects from different metals. And because it is an excellent electrical insulator, soapstone is used in making electric panels. Of course, one of the most popular uses of soapstone has been to make artistic carvings. Unlike marble (pp. 100–101) and granite, which require some heavy-duty tools to carve, all a person needs is a simple knife, some sandpaper, and some artistic talent to create a soapstone masterpiece!

FACTS

ROCK TYPE: metamorphic—low-grade regional metamorphism

MAIN COLORS: white, green, brown

TEXTURE: microscopic crystals

STRUCTURE: foliated, similar to a schist, may be flaky

HARDNESS: 1

COMMON MINERALS: talc, chlorite

CHINESE SOAPSTONE SEAL

MANY DIFFERENT CULTURES, INCLUDING THE **INUIT** IN CANADA, THE PEOPLE OF **CHINA,** AND **ANCIENT EGYPTIANS,** USED **SOAPSTONE** TO CARVE **FIGURES AND ORNAMENTS,** SOME OF WHICH ARE **7,500 YEARS OLD.**

CHINESE
SOAPSTONE
CARVING

Bet You
Didn't
Know!

"Artistic-grade" soapstone, which is used for making sculptures, decorative carvings, and some cookware, is softer than the "architectural-grade" soapstone that is used for making countertops, flooring, and woodstoves. The difference between the two types has to do with the amount of talc found in the stone. Since talc is the softest mineral, the more talc that is present, the softer the soapstone is!

ONE MAGNIFICENT
METAMORPHIC
MOUNTAIN

The state of New Hampshire is home to some of the most fantastic natural scenery in the eastern United States. Near the top of the list is Mount Washington, in the White Mountains, the highest mountain peak in New England, the northeastern region of the United States. At 6,288 feet (1,917 m) above sea level, Mount Washington attracts thousands of visitors from all over the world who hike, bike, drive, and even take a specially designed railroad train to the summit to check out the incredible view. Many visitors are surprised to learn is that this magnificent mountain is not made of granite, which is the official state rock of New Hampshire. (New Hampshire's official state nickname is the Granite State.) It turns out that Mount Washington is a massive mountain of schist (p. 107)!

The story of the rocks of Mount Washington begins almost 600 million years ago, before the start of the Cambrian period. At this point in time, there was no Mount Washington, and the continents of the world were in very different positions from where they are today. This area was covered by a large inland sea into which rivers were carrying layer after layer of sediment that had eroded from the surrounding highlands. Over time, these sediments hardened to form the sandstone and shale that today is known as the Littleton Formation. But that's only the first part of the story!

About 380 million years ago, there was a massive collision of tectonic plates that not only formed the White Mountain range (which at the time was probably as high as the Himalaya are today) but also brought about a big change to the rocks of the Littleton Formation. Heat and pressure from one tectonic plate grinding under another cooked and squeezed the sedimentary rocks, turning them into metamorphic schist (p. 107) and quartzite

IN **1934,** A **WIND** SPEED OF **231 MILES AN HOUR** (372 KM/H) WAS RECORDED ON **THE TOP OF MOUNT WASHINGTON.** IT WAS THE WORLD **RECORD** FOR MORE THAN **60 YEARS!**

MT. WASHINGTON SUMMIT
6,288 FT
1,917 M

MOUNT WASHINGTON

(pp. 102–103). Of course, as all this squeezing and heating was happening these rocks were deep under Earth's surface. But over the past few hundred million years or so, erosion exposed the rocks so that they are now at the surface.

The final sculpting of Mount Washington began a little less than two million years ago, with the beginning of the "Great Ice Age." Glaciers covered the mountain, and, as the rivers of ice flowed, they carved out huge circular depressions near the top of the rocks and valleys down the sides. When the ice finally melted, about 10,000 years ago, it left a barren landscape filled with jagged rocks and one magnificent mountain that people love to climb!

WHY IS **NEW HAMPSHIRE** CALLED THE **GRANITE STATE?**

Less than half of the bedrock found in New Hampshire is made of granite, so why is it called the Granite State? Granite was, and still is, an important economic resource for the people who live there. In the late 18th and early 19th centuries, there was a construction boom in places like New York City and Boston, Massachusetts, and architects and builders loved to use granite in their designs. It just so happens that millions of years of erosion exposed several large deposits of granite in New Hampshire, so enterprising individuals started opening rock quarries from which thousands of tons of granite were mined each year. Today, many of the older quarries have been abandoned, but there are still some active granite quarries in the Granite State—making it true to its name!

When it comes to tall peaks in the northeastern United States, Mount Washington has a lot of company. The five highest mountains in New England are named for the first five presidents of the United States. They are all part of the Presidential Range, which makes up a large section of the White Mountains of New Hampshire.

CHAPTER SIX

REASSEMBLED ROCKS

SEDIMENTARY ROCKS

Sedimentary rocks form when the remains of other rocks join back together again. Sometimes the pieces form thick layers, like they do in the Moenkopi rock formation in the southwestern part of the United States.

DIGGING ON SEDIMENTARY ROCKS

SANDSTONE

Sedimentary rocks make up less than 10 percent of the volume of Earth's crust, yet they are the most common type of rock found on the continents. There's a really good reason for this. Sedimentary rocks form at or near the surface, so they basically form a covering on top of the igneous and metamorphic rocks that make up the underlying crust below.

Loose pieces of sediment and chunks of rocks are called grains, which bond together and harden through a process called lithification to form clastic sedimentary rocks—rocks that are made from pieces or "clasts" of other rocks and minerals after they have been exposed to the forces of weathering. These include rocks such as conglomerate (p. 127), sandstone (p. 127), and shale (p. 127). There are several ways that sediments can become lithified. Large pieces of sediment such as cobbles, gravel, and sand usually get stuck together by some type of cement, in a similar way that people mix up a batch of concrete. In this case, instead of coming from a bag, the cement is carried into the sediment by mineral-rich water flowing through it. The minerals fill up the spaces between the grains, turning them into a solid mass. Three of the most common types of mineral cements are silica, calcite, and iron oxide, which usually turns the rock a rusty red color. Because of the way the sediments are deposited, sedimentary rocks often have distinct layers called beds in them, but in some cases the layering will not be visible. When sedimentary rocks form thick deposits with no layers in them, they are called massive.

Clastic sedimentary rocks can also harden without cement if the particles are squeezed together under enough pressure. This happens when layers of fine-grained sediments like silt and clay get buried deep under piles of other sediment, the

PORTLAND CEMENT, WHICH IS THE **MOST COMMON TYPE** OF CEMENT THAT IS USED TO MAKE CONCRETE, IS **MADE FROM LIMESTONE** THAT HAS BEEN CRUSHED, **MIXED WITH CLAY,** AND HEATED TO A HIGH TEMPERATURE.

A CHALK FORMATION IN EGYPT'S WHITE DESERT

Bet You Didn't Know!

There are some sedimentary rocks that don't neatly fit into either the clastic or chemical groups. Known as organic sedimentary rocks, these are rocks that form from the accumulation of material that was once part of a living thing. Coal (pp. 15, 256–257), for example, comes from plant material, and chalk (pp. 122–123) and coquina (p. 121) are made up of shells of dead sea creatures that have piled up over time.

Sedimentary rocks come in a wide range of textures and compositions, but they all form from the leftovers of other rocks.

SHALE

GRAPPLING WITH
GRAIN SIZE

When you hear the word "sand," what's the first thing you think of? Building castles at the beach, perhaps? How about "clay"? That's the stuff you make pottery from, right? And what about a "boulder"? That's a great big rock that's usually too big to move. We use terms like "sand," "clay," and "boulder" in our daily lives, but most people don't realize that, to a geologist, each of these words has a very specific meaning used in describing different types of clastic sedimentary rocks. And a scientist named Chester Keeler Wentworth is to thank for it!

In the 1920s, Wentworth defined each of these terms, as well as about a dozen more, giving exact sizes for each. A boulder, for example, is any rock fragment bigger than 10 inches (25.6 cm) in diameter, while sand can be a particle that ranges from .002 to .08 inch (.062–2 mm) in size. Clay particles are so tiny that you need a microscope to see them, and they're defined as any grain smaller than .00016 inch (.004 mm) in size. This scale, which is called the Wentworth grain size scale in his honor, helped to revolutionize the classification of sedimentary rocks.

weight of which is enough to get the tiny grains to stick together forming rocks such as shale (p. 127) and mudstone (p. 126).

Rocks that form from the evaporation of mineral-rich water are called chemical sedimentary rocks. Instead of being made up of pieces of sediment, these rocks contain mineral crystals that precipitate, or come out of, the water and grow together as the water slowly dries up. Sometimes they form evaporite rocks, such as rock salt (pp. 186–187) and gypsum rock (pp. 200–201), right at the surface. In other cases, the minerals precipitate out of seawater and get deposited on the bottom of the ocean, often forming an ooze that eventually dries out and hardens. This is how some limestone (pp. 120–121) forms.

ROCKIN' COTTON CASTLES

When visitors first arrive in the town of Pamukkale in Denizli Province in southwestern Turkey, some don't quite know what to make of it. From a distance it almost looks like the ground is coated in a thick blanket of fluffy white cotton. (The Turkish name Pamukkale actually means "cotton castle.") But when you take a closer look, you see that the "cotton" is really made from deposits of a rock called travertine, which forms from minerals that bubble up from hot springs coming out of the ground. As the water flows down over a series of low cliffs, the minerals that make the travertine come out of the water and form natural terraces that look like they are part of some type of fantastic sculpture built by an alien race.

Travertine is classified as a chemical sedimentary rock that is related to limestone (pp. 120–121), and it forms only in a few special environments. In the case of Pamukkale, the process begins when groundwater that has been heated by magma deep underground flows up through the surrounding limestone, dissolving the calcite (pp. 190–191) from which it is made. The hot water, which is now rich in calcium carbonate, eventually flows out of the ground in a series of 17 individual hot springs that usually range in temperature from 95 to 140°F (35–60°C). The water moves down the low cliffs, carbon dioxide gas (the same stuff that makes soda fizz) bubbles out of it, and the calcium carbonate that it carries begins to precipitate. At first, the newly formed travertine is soft, almost like jelly, but over time it hardens into solid rock.

As with many natural wonders in the world today, the travertine terraces at Pamukkale have been impacted by the thousands of tourists who would come to bathe in the mineral-rich waters found in some of the natural pools. This, coupled with the

TRAVERTINE DOESN'T JUST FORM AT **HOT SPRINGS.** IT CAN ALSO BE FOUND IN **CAVES** WHEN WATER RICH IN **CALCIUM CARBONATE** DRIPS DOWN OVER THE **ROCKS.**

Travertine terraces can be found in several different locations around the world, including Yellowstone National Park in the United States. The Mammoth Hot Springs terraces found there are formed by mineral-rich water that is thought to be heated by the same deep magma pool that fuels the park's famous geysers.

The travertine terraces located in Pamukkale, Turkey, look like fluffy layers of cotton and are formed as minerals grow from hot springs flowing out of the earth.

BUILDING WITH **TRAVERTINE**

Not only does travertine make unbelievable land-scapes in the natural world, but—over the centuries—people have used it to create some pretty impressive structures, too! One of the most famous is the Colosseum in Rome, Italy. Shaped like a giant oval and big enough to fit a modern-day soccer field inside, the Colosseum is 12 stories high and when it was in regular use could seat 50,000 people. Construction began in A.D. 72 and was completed in less than 10 years. The Colosseum was one of the first major structures in Rome made with concrete, but most of the large stones used for the piers and arch-ways, which are still standing today, are made from travertine. Travertine wasn't just used in ancient structures, though. The Getty Center in Los Angeles contains approximately 1.2 million square feet (111,000 sq m) of both polished and unpolished Italian travertine panels.

construction of hotels above the springs, contami-nated the water and threatened to destroy some of the natural travertine formations. Fortunately, in 1988 the site was named a UNESCO World Heritage Site, a designation that put restrictions on what people can do there. In addition, some of the hotels that were built along the springs were demolished and one of the main roads was removed so that people now have access to only a few of the pools. Hopefully this will help to preserve the "cotton castles" for many years to come.

LIMESTONE

Limestone is one of the most common sedimentary rocks on Earth's surface. Most of the limestone that we find exposed on the continents today formed underwater in warm shallow seas that were home to lots of sea creatures such as mollusks, corals, brachiopods, and especially tiny single-celled organisms called foraminifera. Even though these critters looked very different from one another, they had one thing in common: They all made hard shells to protect themselves, and these shells were all made from calcium carbonate, which they extracted from the seawater. Another name for calcium carbonate is lime, and that's where the "lime" in "limestone" comes from.

When the creatures died, their shells sank to the bottom of the sea and began to pile up. Quite often the shells broke up and partially dissolved, but then over time the mineral calcite would precipitate back out of the water forming layers of limestone. In some cases, instead of forming distinct layers, the limestone formed large, thick deposits called massive limestone. Since this type of limestone was formed by living creatures, it's known as biogenic limestone. It has tiny calcite grains that usually require a microscope to see but sometimes are large enough to see with the naked eye. In many cases, fossils of some of the creatures are in the limestone!

Limestone also forms as a result of tiny crystals of calcite forming in the water when it has a lot of calcium carbonate dissolved in it. You can sometimes see this type of limestone forming today in caves, near hot springs, and in salty lakes that have high evaporation rates.

Over the years, people have come up with many ways to use limestone. It's mined to create building stones and crushed to make lime that farmers use to add nutrients to the soil. It's the main ingredient in most types of building cement and is also used in the steelmaking process to remove unwanted materials from the molten metal. Even though it is quite common, limestone is an extremely valuable resource, and we have billions of sea creatures to thank for it!

FACTS

ROCK TYPE: sedimentary—chemical or biogenic

MAIN COLORS: light to dark gray, white, yellow, occasionally black

TEXTURE: granular, often with small crystals; in some cases, grains of calcite are visible to the naked eye; some varieties contain shell fragments and fossils

STRUCTURE: usually in massive beds but layering is often present

HARDNESS: 3–4

COMMON MINERALS: calcite, dolomite, aragonite

SOME LIMESTONES CONTAIN **TINY SPHERES** CALLED **OOLITES** THAT FORM WHEN **CALCIUM CARBONATE** GROWS AROUND **SAND GRAINS** THAT ARE ROLLING AROUND ON THE **BOTTOM OF THE SEA.**

Limestone is usually white or light gray in color and can be layered or massive.

COQUINA

Bet You Didn't Know!

Coquina is a special type of limestone that is made from smashed-up seashells. Sometimes geologists call this a bioclastic rock because, like other clastic sedimentary rocks, it's made from fragments that have been cemented together again. But in this case, instead of being grains of sediment, the fragments are from the shells of living things.

The White Cliffs of Dover are made from chalk that came from the skeletons of tiny sea creatures that lived millions of years ago.

Bet You Didn't Know!

At one time, people had only natural chalk to use for writing or drawing, but over time things have changed. The chalk that is used today to draw on sidewalks and chalkboards is manufactured by mixing calcium carbonate and gypsum powder with clay—and it's also often dyed to make different colors.

CHALK

There is no mistaking the White Cliffs of Dover! Erupting like a great white wall from out of the sea, they are one of the great natural landscapes in the world. Standing more than 300 feet (100 m) high, they form a 10-mile (16-km)-long barrier along the coastline of England. On a clear day, they can be seen all the way from France—on the other side of the English Channel—some 25 miles (40 km) away. What makes these massive cliffs even more spectacular is that they are made of chalk, a supersoft sedimentary rock that got its start with some of the tiniest sea creatures to live on the planet.

Technically speaking, chalk is a type of limestone (pp. 120–121), but the way it forms is so different that it deserves special recognition as a unique sedimentary rock. Before you can have a chunk of chalk, you need a relatively deep ocean environment that does not have a lot of sand, silt, or clay flowing into it. Floating around in the water are billions of microscopic organisms called foraminifera and coccoliths, both of which have hard parts made of calcium carbonate, also known as lime. As they die, the hard parts rain down on the seafloor forming a thick layer of sediment that slowly turns into a limey ooze. As more and more material builds up, the ooze at the bottom of the pile gets compacted and turns into a solid rock that we call chalk.

Today, there isn't much chalk forming in the oceans. But somewhere around 100 million years ago, during the Cretaceous period, things were a lot different. Back then the atmosphere was much warmer, and most of the polar ice had melted, causing sea levels to rise over 300 feet (100 m). As a result, much of the area that is now Western Europe was underwater, and large chalk deposits were forming at the bottom of the sea. Over time, tectonic forces raised the rocks above sea level, exposing the chalk to erosion, and creating the pearly white cliffs we see today.

FACTS

ROCK TYPE: sedimentary—biological (organic)

MAIN COLORS: white, yellow, occasionally reddish

TEXTURE: very fine grains, often with tiny microscopic fossils

STRUCTURE: usually in massive beds, but layering is often present

HARDNESS: less than 3; easily crumbles

COMMON MINERALS: calcite, clay minerals

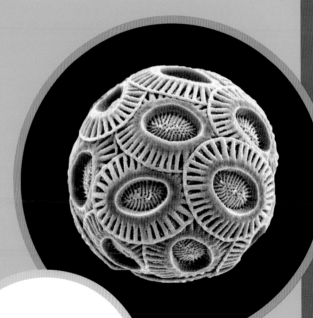

IMAGE OF COCCOLITHS MADE BY A SCANNING ELECTRON MICROSCOPE

THE **CRETACEOUS** TIME PERIOD FROM **145** TO **66 MILLION YEARS AGO** GETS ITS NAME FROM THE **LATIN WORD CRETA,** WHICH MEANS **"CHALK."**

TIME TO MAKE THE BEDS

When exploring sedimentary rocks out in the field, one of the first things that jumps out at you is the fact that they are made up of different layers, just like a fancy cake (only without the whipped cream)! When you stop and think about it, this makes a lot of sense, because when sediments are deposited by rivers and streams on lake bottoms or the ocean floor, they tend to pile up on top of each another. Each layer in a sedimentary rock is called a bed, and geologists spend a lot of time studying them. Why? Because beds can tell us a great deal about the environment that the rocks formed in. For example, geologists can go to a place such as a delta where sediment is being deposited along the edge of the ocean and observe the way the individual beds form. They can then look at the beds preserved in a sedimentary rock. If they see the same features that they saw in the modern sediment, they can get an idea of the type of flow that happened when the sediments in the rock were being deposited. Basically, they use what is happening today and show that it is the same process that happened in the past.

One of the most important things that geologists look for in the beds is sedimentary structures. These are features such as ripple marks and mud cracks that formed as the sediment was being deposited by water. If you have ever been to a sandy beach where the waves aren't too large, you've probably seen ripple marks. When the water on the beach sloshes back and forth it makes little ridges in the sand called ripples. If you look closely, you'll see that one side of the ripple is usually steeper than the other because of the direction the water was moving in. Over time, sediment covers the ripples, which preserves them,

A **GRADED BED** IS A **LAYER** IN THE **SEDIMENT** THAT STARTS WITH **LARGE PIECES** OF SEDIMENT AT THE **BOTTOM** OF THE BED AND GRADUALLY CHANGES TO **FINER SEDIMENT** AT THE **TOP.** IT FORMS ON THE **EDGE OF A LAKE OR OCEAN** WHEN A LARGE MASS OF **MIXED SEDIMENT** BEING CARRIED HAS TIME TO **SLOWLY SETTLE** OUT.

MUD CRACKS IN A SEDIMENTARY ROCK

AN UPSIDE-DOWN GRADED BED

Bet You Didn't Know!

As pieces of sediment get carried in streams, the rolling and tumbling tends to knock off their sharp edges. The farther a rock gets transported, the rounder and more spherical it gets. If you find a sedimentary rock with nothing but round grains in it then there is a good chance that it was transported a fairly long distance over a long period of time.

Features such as ripple marks preserved in sedimentary rocks give geologists important clues about the environment that the rocks formed in.

and eventually the sediment becomes a sedimentary rock. Seeing ripple marks in sedimentary rocks tells a geologist which direction the water came from.

Mud cracks form in areas along the edge of streams, ponds, or tidal marshes that have lots of fine-grained silt and clay. Usually these areas stay wet, but when they dry out the surface starts to crack making a distinctive pattern. Over time, additional sediment covers the mud cracks preserving them. Finding mud cracks in sedimentary rocks tells geologists about the way the environment of an area changed over time and helps them work out the geologic history of the area.

SORTING IT ALL OUT

In most cases, the faster the water flows in a river, the larger the pieces of sediment it can transport. When a river enters a lake or the ocean, it loses energy and can no longer carry the same amount of sediment, so the sediment gets deposited. The first sediment particles to be deposited are the largest, heaviest chunks, like gravel and coarse sand. The water still has enough energy to carry the fine sand and silt, so these particles get deposited in deeper water. The last particles to be deposited are the supersmall clay particles, which wind up in the deepest water. This process of depositing sediments is called sorting, and it creates one of the most important features found in sedimentary rocks. If a rock is well sorted, like a sandstone (p. 127), then all the grains in it are pretty much the same size and they have been deposited by a steady flow of water over a long period of time. A poorly sorted rock would be something like a conglomerate (p. 127), which is a jumbled mishmash of different sediments all mixed together. These usually form as a result of a sudden flood of water mixing the different sediment types together.

A RANGE OF ROCKS

By far, the most common sedimentary rocks found on Earth's surface are made from grains of sediment that have been either cemented or squeezed together. These clastic rocks are given different names based on the size of their grains, how well the grains are sorted, and in some cases, the type of mineral that makes up most of the rock. Here's a rundown of five of the most common clastic sedimentary rocks.

BRECCIA

A sedimentary breccia is one mixed-up rock! It's pretty much a jumbled mass of different-size sediments that were basically dumped in place and then, over time, hardened into stone. To be called a breccia, at least half the particles must be larger than .08 inch (2 mm) across, and these larger fragments must be angular rather than rounded. The sharp angles on the fragments means that they were transported only a short distance and, since there is little or no sorting, they were deposited quickly, usually by some type of flood or underwater landslide.

MUDSTONE

If you think that a mudstone looks a lot like a shale, you're right. In fact, many geologists consider mudstone to be a type of shale. They are both made of supertiny silt and clay particles that are smaller than .0025 inch (.062 mm) in size and they are both well sorted. The main difference between the two is that shales have very distinct beds that allow them to split easily, while mudstones tend to form in massive layers that can be several feet thick! They often have calcium carbonate mixed in them and lots of fossils. They frequently have structures such as mud cracks on top of some of the layers showing that the environment that they formed in dried out from time to time.

SHALE

Shale is the most common clastic sedimentary rock found in Earth's crust. It is an important source of rock particles and clay minerals for use in making concrete, tiles, and bricks. Shale is a well-sorted fine-grained rock made from tiny silt and clay particles that were carried by slow-moving currents of water and deposited in quiet water environments where they had time to slowly fall to the bottom. It is quite common to find layers of shale alternating with layers of limestone (pp. 120–121) and sandstone. Shales usually have very thin beds and split easily into layers. They also often (but not always) have fossils in them. Shale is most often gray or black but can come in many different colors including red, brown, and even green.

SANDSTONE

True to its name, sandstone is a stone that is made up of mostly sand-size particles that range from .0025 to .08 inch (.062–2.0 mm) in diameter. Sandstones are well-sorted, with most of the particles being the same size. They usually have distinctive layers or beds indicating that they were deposited close to shore by fast-flowing water or by wind in sand dunes. Sandstones are the second most common sedimentary rocks found on the continents, and even though the sand grains are cemented together, there is usually a great deal of empty pore space between the grains. Because of this, sandstones often hold fluids like water, oil, or natural gas in their pore spaces. If a sandstone is made of just sand, it is often called an arenite, while a sandstone with more silt and clay in it is called a wacke.

CONGLOMERATE

Like a breccia, a conglomerate is a rock that is usually poorly sorted with a wide range of grain sizes, some of which can be quite large. Conglomerates with pebbles and cobbles in them are quite common, and there are even such things as boulder conglomerates, where the largest chunks of sediment are actual boulders! The main feature that separates a conglomerate from a breccia is that the particles in conglomerates are all rounded. This means that they have been transported a relatively long distance from their source, giving the water source time to round out the rock. The fact that the rock is usually a mixed-up mass of sediment sizes means that the grains were deposited quickly by fast-flowing water.

DISCOVERING DINOSAURS

While dinosaur fossils have been found in sedimentary rocks on every continent, they are not spread evenly around the planet. First, you need to have rocks that are the correct age—in this case, from the Mesozoic era (about 252 to 66 million years ago). Next, the rocks have to be exposed near the surface, which means that they need to be in an area that has had quite a bit of erosion. Finally, it's best to have a location that is not near cities where there is a lot of pavement or areas that are covered with grass or trees because these things tend to cover up the fossils. For these reasons, many of the best dinosaur discoveries come from areas that have dry climates such as the Gobi desert in Mongolia, the Patagonian desert in Argentina, and the badlands of Alberta, Canada, and Montana and North Dakota in the United States.

A lot of what paleontologists believe about dinosaurs has changed over the past 50 years. Scientists once deemed dinos cold-blooded, lumbering lizards that became extinct because they were not able to compete with us smarter, warm-blooded mammals. These days, however, a totally new picture of dinosaurs has emerged—thanks to lots of new fossil evidence found in sedimentary rocks! Today, paleontologists are leaning toward the idea that many dinosaurs had the ability to control their body temperatures at least a little. Many paleontologists also believe that theropod dinosaurs like *T. rex* and *Microraptor* were more closely related to birds than to lizards.

Ever since the discovery of *Archaeopteryx* fossils in Germany in 1861, some scientists thought there might be a connection between birds and dinosaurs. *Archaeopteryx* has been called a missing link between birds and dinosaurs because it had sharp teeth and a long bony tail like a lizard but also wings with flight

JUVENILE
T. REX

THE WORD **"DINOSAUR"** WAS FIRST USED BY ENGLISH SCIENTIST **SIR RICHARD OWEN** IN 1841. IT COMES FROM TWO **GREEK WORDS** THAT TOGETHER MEAN **"TERRIBLE LIZARD."**

About 100 million years ago, Dinosaur Provincial Park (above), in the badlands of Alberta, Canada, was a swampy floodplain with many sediment-choked rivers flowing through it, making it a perfect environment for many different dinosaurs to call home. When they died, their bones were quickly buried and preserved. After millions of years of erosion, the sandstone and shale with the fossils have become exposed. Over the past century, thousands of individual fossils from more than 30 different dinosaur species have been found!

ARCHAEOPTERYX

DOOMSDAY FOR THE **DINOS!**

Paleontologists had come up with a few different theories to explain why dinosaurs suddenly died out at the end of the Cretaceous period. Some suggested that it was due to a large increase in volcanic activity, and others pinned the blame on the rise of other animal groups. Then, in 1980, a geologist named Walter Alvarez suggested a totally different possibility.

Alvarez discovered a thin layer of clay dating to the exact time that the dinosaurs, along with hundreds of thousands of other life-forms, died out. The clay was rich in an element called iridium, which is rare on Earth but common in meteorites. Alvarez and his father, Nobel Prize–winning physicist Luis Alvarez, proposed that the mass extinction was caused by a huge asteroid hitting Earth, which set off a chain of events that changed Earth's climate so drastically that it wiped out most of the life-forms on the planet. The strongest evidence came with the discovery of a giant crater buried beneath sediments in the Gulf of Mexico that formed at the same time of the asteroid impact. Paleontologists finally had a solid theory to explain what caused dinosaurs to go extinct! Today some paleontologists believe dinosaurs were already declining in number by the time of the asteroid strike—making it even harder for them to recover from the environmental changes caused by the catastrophic event.

feathers like a bird. Over the past two decades, a lot more fossil evidence supporting this connection has been discovered. Many fossils clearly show feathers on a wide range of theropod dinosaurs, including *Velociraptor*.

Like modern-day birds, some dinosaurs also guarded their nests and protected their eggs. Also, some dinosaurs, like *Allosaurus*, had hollow bones, just like birds, and scans of the fossilized skulls of theropod dinosaurs show that they had the same highly developed brains needed to control flight. Next time you see a bird flying past your window, you'll know it's a relative of dinosaurs!

THE CHICXULUB CRATER UNDER THE WATERS OF THE GULF OF MEXICO

CHERT & FLINT

You may be wondering why we have two rocks listed together here. The answer is simple—they are basically the same rock! Even though many people call them by separate names, flint is really a type of chert that's usually black, dark gray, or brown, while chert is usually white or light gray. Both are composed of microcrystalline silica—a complicated way of saying supertiny grains of quartz (pp. 206–207). Chert can be found as thick beds or as small rounded lumps in limestone (pp. 120–121) and in layers of sediment that have not yet turned to stone.

There are several ways that chert can form. The first method is very similar to the way chalk (pp. 122–123) forms. Skeletons of tiny marine organisms called diatoms and radiolarians build up on the seafloor and then turn into a jellylike ooze that hardens and recrystallizes. The difference between chalk and chert is that the skeletons of the organisms that make chert are all composed of silica instead of calcium carbonate. In many cases you can find fossils of their skeletons in the chert, and, as you might expect, chert and chalk are often found together.

The second way chert forms is through a chemical process: Groundwater containing dissolved silica flows through limestone and replaces some of the mineral calcite (pp. 190–191). This creates a rounded lump of chert called a nodule.

Flint was a very important resource for Stone Age people because they used it for catching and cooking their dinner. Because it is made of almost pure quartz, flint is very hard. When it breaks, it can also create an incredibly sharp edge. These two properties made it an ideal material for making cutting tools and weapons. The people also discovered that if you strike a chunk of pyrite (pp. 160–161) that has lots of iron in it with flint it would make a big spark. This allowed people to start fires anytime they wanted, such as when it was time to cook the dinner that they had caught using their flint weapons.

FACTS

ROCK TYPE: sedimentary—chemical or biogenic

MAIN COLORS: light to dark gray, brown, red, white, yellow, black

TEXTURE: extremely fine grained, almost glassy

STRUCTURE: thick beds or in nodules; breaks with shell-like conchoidal fracture

HARDNESS: 7

COMMON MINERALS: quartz

FLINT

ANCIENT EGYPTIANS CRAFTED BRACELETS OUT OF FLINT.

ANCIENT CHERT CUTTING TOOLS

CHERT

Bet You Didn't Know!

People have come up with lots of different names for chert depending on its color and where it is found. Flint that forms nodules comes in dark colors, while jasper is chert that has been stained red by iron oxide. When chert is translucent, meaning light can pass through it, it is usually called chalcedony. There's also the light gray metamorphosed form of chert found in the mountains of Arkansas, U.S.A., called novaculite and a colorful variety of microcrystalline quartz from Australia called mookaite!

THE GRANDEST
OF ALL CANYONS

T**he Grand Canyon is one of the most spectacular geologic environments on the planet. Located in the northwestern part of the U.S. state of Arizona, the Grand Canyon is a deep gorge** cut into the earth by the Colorado River that stretches more than 275 miles (443 km). On average, it's one mile (1.6 km) deep and in places 18 miles (29 km) wide. Geologists still aren't sure exactly how old the Grand Canyon is, but we know that it is at least six million years old and may have started forming as much as 70 million years ago. That's when tectonic forces inside Earth lifted the Colorado Plateau, allowing the Colorado River to begin its slow, steady cut through millions of years of Earth's history.

At the lowest levels of the canyon you will find Vishnu schist (p. 107), the metamorphic rocks that form the basement of the canyon and were once the roots of giant mountains that have long since eroded away. In the schist are intrusions of granite (pp. 78–79) that formed when magma flowed up through the rocks and cooled before reaching Earth's surface.

Resting on top of these metamorphic and igneous rocks are thousands of feet of sedimentary rocks that record the geologic history of the area, starting with a time it was covered by the sea. In the middle of the canyon are alternating layers of sandstone (p. 127), shale (p. 127), and limestone (pp. 120–121), which show how the depth of the water changed as sediments filled the area. Fossils of corals and other warmwater animals tell us that the area had a tropical climate. As you near the top of the canyon, the rocks again change, showing that the area was later above sea level and home to the floodplain of a large, meandering river. Finally, around 275 million years ago, the area became so dry that it was covered in sand dunes forming the sandstones that are found near the

THE GRAND CANYON

ARIZONA

While most of the rocks found in the Grand Canyon were deposited under water, the Coconino Sandstone near the top of the canyon was deposited by wind. Geologists figured this out by looking at the structures and fossils in the rock. The Coconino Sandstone has lots of thin layers of sand at a steep angle called crossbedding, which is usually found in sand dunes. There are also fossil tracks of millipedes and scorpions that lived in a desert environment.

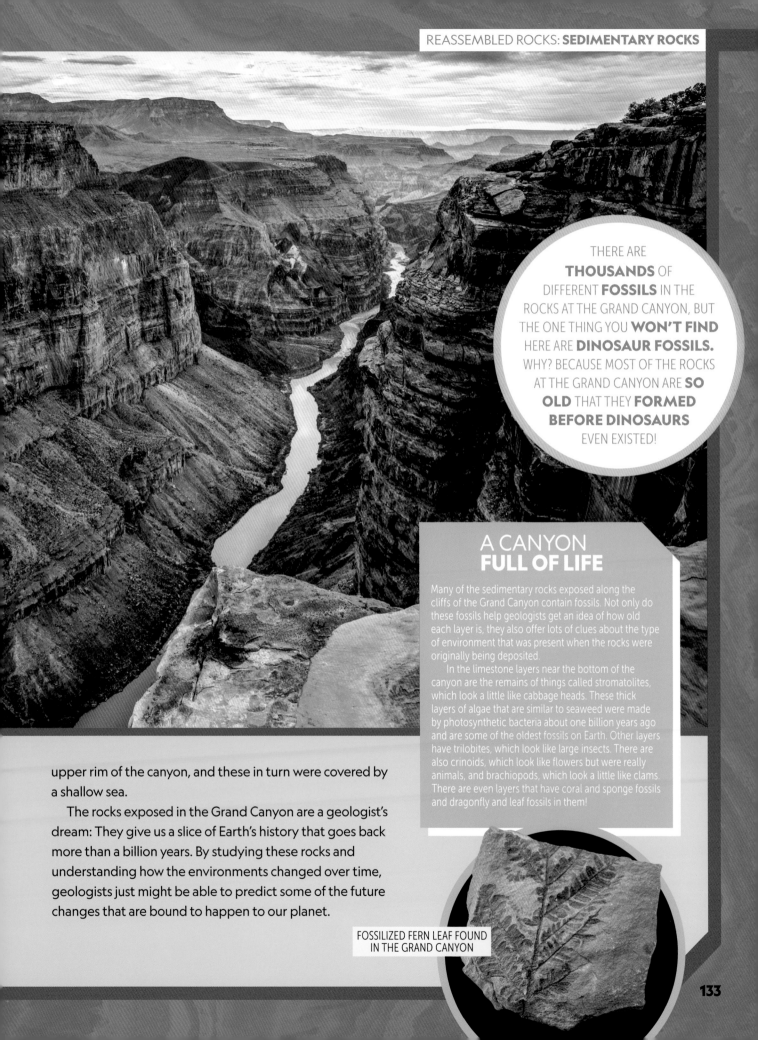

THERE ARE **THOUSANDS** OF DIFFERENT **FOSSILS** IN THE ROCKS AT THE GRAND CANYON, BUT THE ONE THING YOU **WON'T FIND** HERE ARE **DINOSAUR FOSSILS.** WHY? BECAUSE MOST OF THE ROCKS AT THE GRAND CANYON ARE **SO OLD** THAT THEY **FORMED BEFORE DINOSAURS** EVEN EXISTED!

A CANYON
FULL OF LIFE

Many of the sedimentary rocks exposed along the cliffs of the Grand Canyon contain fossils. Not only do these fossils help geologists get an idea of how old each layer is, they also offer lots of clues about the type of environment that was present when the rocks were originally being deposited.

In the limestone layers near the bottom of the canyon are the remains of things called stromatolites, which look a little like cabbage heads. These thick layers of algae that are similar to seaweed were made by photosynthetic bacteria about one billion years ago and are some of the oldest fossils on Earth. Other layers have trilobites, which look like large insects. There are also crinoids, which look like flowers but were really animals, and brachiopods, which look a little like clams. There are even layers that have coral and sponge fossils and dragonfly and leaf fossils in them!

upper rim of the canyon, and these in turn were covered by a shallow sea.

The rocks exposed in the Grand Canyon are a geologist's dream: They give us a slice of Earth's history that goes back more than a billion years. By studying these rocks and understanding how the environments changed over time, geologists just might be able to predict some of the future changes that are bound to happen to our planet.

FOSSILIZED FERN LEAF FOUND IN THE GRAND CANYON

133

CHAPTER SEVEN

THE **BUILDING BLOCKS** OF **ROCKS**

Minerals form in different ways and come in many shapes and colors. Some, like the amethyst crystals in this geode, look spectacular, while others are pretty dull. But whether sparkly or subdued, it's minerals that make rocks possible.

BREAKING DOWN SOME MINERAL PROPERTIES

Here's a little experiment for you to try the next time you take a trip to a sandy beach: Bring along a saltshaker, a magnifying glass, and a paper plate. Use a marker to draw a line down the middle of the plate. Sprinkle some salt on one side of the plate and some sand on the other. Then look at both with the magnifying glass. Notice any differences? Chances are the edges of the sand grains will be rounded and cracked in a random pattern, while the salt crystals will mostly have square edges. This difference comes from a mineral property called cleavage, and it's really helpful in identifying some minerals.

When atoms and molecules join together to form crystals, there will often be places where the bonds that hold them together are not as strong. These zones of weakness are called cleavage planes, and they often control the way the mineral breaks. Minerals like muscovite (p. 217) and biotite (p. 217) have perfect cleavage in one direction, so you can peel the mineral apart into thin sheets. Halite (pp. 186–187), which is the mineral name for table salt, has cleavage in three directions at right angles to each other, so it breaks apart into cubes.

Some minerals, such as quartz (pp. 206–207), which is a common mineral found in beach sand, don't have these zones of

QUARTZ

BECAUSE OF ITS **PERFECT CLEAVAGE** AND **HIGH MELTING POINT,** SHEETS OF **MUSCOVITE** (P. 217) ARE **USED FOR WINDOWS** TO SEE INSIDE SPECIAL FURNACES USED FOR **HEATING METALS.**

MUSCOVITE

MOHS HARDNESS SCALE

1 → INCREASING HARDNESS → **10**

1 TALC

2 GYPSUM

3 CALCITE

4 FLUORITE

5 APATITE

6 FELDSPAR

7 QUARTZ

8 TOPAZ

9 CORUNDUM

10 DIAMOND

weakness, so they don't have cleavage. Instead, when you hit them, they fracture in a random pattern, like a piece of glass. This is called conchoidal fracture, and the split edges look a little like the curved piece of a seashell.

Another mineral property that is related to cleavage is hardness, which is a measure of how tough a mineral is to scratch. Some minerals, such as gypsum (pp. 200–201), are so soft that you can scratch them with your fingernail. Others, including the feldspars (pp. 212–213), are so hard that they can scratch a piece of iron. When it comes to identifying minerals, hardness is really helpful. Whether it is a large crystal or a tiny grain, the hardness of a particular mineral always stays the same, although some minerals have a different hardness depending on the direction that you scratch them.

MR. MOHS
MAKES A SCALE

Friedrich Mohs was a German geologist who was fascinated with minerals and their different properties. Back in the 1800s, the most common properties that people used to identify minerals were their crystal shape and color. This created a problem, though, because minerals don't always form perfect crystals, and sometimes a single mineral can come in lots of different colors.

Working with several large collections, he realized that some minerals were super soft and could be easily scratched, while others were almost impossible to scratch, even with a steel knife. He began testing and classifying (or sorting) out different minerals based on how easy they were to scratch, using a scale of 1 to 10. Talc (pp. 220–221), which is the softest mineral, was classified as 1, and the hardest, diamond (pp. 152–153), was given the number 10. He then selected eight other minerals to serve as reference points on the scale. This system is called the Mohs scale of mineral hardness, and it is one of the most important tools used by geologists for identifying minerals.

Bet You Didn't Know! Each mineral in the Mohs scale can scratch any other mineral with the same or a lower number but can't scratch any mineral with a higher number. This means that quartz at number 7 can scratch feldspar at number 6, but neither can scratch topaz at number 8.

WELCOME TO THE CLASS

S cientists love to classify things. Classification simply means placing objects into different groups based on their similar characteristics or properties. There are classification systems for animals, plants, stars, clouds, and, of course, for rocks and minerals. Rocks are classified as being igneous, metamorphic, or sedimentary based on the way they form. When it comes to minerals, the most common classification system involves placing them into different mineral classes based on their chemical makeup.

Minerals that are in the same class share a lot of the same properties, so knowing a mineral's class gives a geologist some idea of how a mineral will behave and the type of environment that it formed in. The following is a rundown of some of the largest mineral classes based on the way the chemical elements join together.

NATIVE ELEMENTS—These minerals are all pure elements and include precious metals such as gold (pp. 142-143) and silver (pp. 146-147).

SULFIDES—These minerals form when the element sulfur combines with a metal. Some examples are pyrite (pp. 160-161) and galena (pp. 158-159), both of which are considered ores (rocks that are mined from the ground to get the metal that they contain).

OXIDES—These minerals make up many of the ores that are mined to get metals like iron and aluminum. They are formed when oxygen combines with a metal. This class includes hematite (p. 174) and corundum (pp. 166-167), from which we get the gems ruby and sapphire.

NOT ONLY DO **SILICATE MINERALS** MAKE UP ABOUT **95 PERCENT** OF **EARTH'S CRUST** AND **UPPER MANTLE,** THEY ARE ALSO FOUND IN **MOON ROCKS, METEORITES, AND ASTEROIDS!**

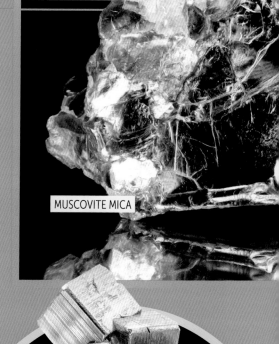

MUSCOVITE MICA

PYRITE

CHRYSOBERYL

One reason that chrysoberyl, pyrite, and muscovite mica look very different from one another is that they are all in different classes of minerals with different chemical compositions.

IN A CLASS BY HIMSELF!

It can be argued that no single person has done more for the science of mineralogy than American scientist James Dwight Dana. Born in 1813, Dana always had a love of the natural world, especially rocks and minerals. At that time, minerals were grouped in many different ways, so it was often difficult to figure out how they all fit together. After graduating from Yale University in 1833, Dana began working on a new, more efficient system for classifying minerals. He decided that he would base his system on the same type of system that the Swedish naturalist Carolus Linnaeus had developed for classifying animals and plants in the 1730s.

In 1837, at the age of 24, Dana published an epic book called *A System of Mineralogy*. Over the next 20 years, Dana worked on making many revisions to the system based on his own work with minerals. By the time the fourth edition was published, in 1854, he had fully developed the classification based on the chemical composition of minerals that geologists still use today.

HALIDES—These are soft minerals that tend to dissolve in water. This class includes the mineral halite (pp. 186–187), which is a main source for table salt.

CARBONATES—All of these minerals are made from a chemical compound made from a carbon atom joined with three oxygen atoms. Carbonates include the minerals calcite (pp. 190–191) and dolomite (pp. 194–195).

PHOSPHATES—Phosphates usually get deposited when hot, mineral-rich water flowing through rocks cools. They also form as the result of chemical weathering of other minerals. The minerals in this class all contain the element phosphorus and four oxygen atoms as part of their makeup. One common phosphate mineral is the gemstone turquoise (pp. 198–199).

SULFATES—These soft minerals form when sulfur and four oxygen atoms combine with another element. Gypsum (pp. 200–201) is a member of this class.

SILICATES—By far the largest mineral class, all of these minerals are made from a chemical structure called the silicon tetrahedron, which is made from a single silicon atom and four oxygen atoms. It is shaped like a tiny pyramid, with an oxygen atom at each corner and the silicon atom in the center. This class includes most rock-forming minerals such as quartz (pp. 206–207), feldspars (pp. 212–213), micas (pp. 216–217), amphiboles (pp. 228–229), and pyroxenes (pp. 224–225).

Bet You Didn't Know!

Individual types of minerals are called species, which is the same term that is used to describe individual types of animals or plants. Sometimes minerals are placed into families based on similar chemistry, and mineral families are divided into groups, each of which has the same general crystal structure.

NATIVE SULFUR

If you like yellow rocks that smell bad and easily crumble when you hit them with a rock hammer, then sulfur is the mineral for you! Sulfur is easy to identify because it has so many distinctive properties that set it apart from most other minerals found in the world. Sulfur is a common element that combines with other chemicals to form minerals such as sulfides and sulfates, but it can also be found as a pure native element.

Native sulfur is often found around active volcanoes and hot springs where it forms from hot liquids and gases that come out of openings in the ground called vents. While it can form crystals of different shapes, sulfur's crystals most commonly look like little pyramids. Most often sulfur is just found as a crust coating other rocks around the vent.

Sulfur is quite soft and can be easily scratched with a penny. It's also very brittle with only slight cleavage, so when it is hit with a rock hammer it usually shatters, sometimes leaving sharp edges. Sulfur also feels light compared with other minerals. How heavy a rock or mineral feels is a property called specific gravity. Sulfur has a specific gravity of only 2, while most rocks and minerals are closer to 5. One property that makes sulfur really stand out is its smell, which is sort of like rotten eggs!

Sulfur is an important industrial chemical and is used in many different products including sulfuric acid, which is used in making car batteries. Sulfur deposits sometimes are found underground near salt domes; it is mined by pumping superhot water down into the sulfur, which causes it to dissolve. The sulfur-rich water is then pumped to the surface and cooled so that the sulfur can be recovered.

FACTS

CLASS: native elements

MAIN COLORS: yellow but may have a green tint

CLEAVAGE: poor; fractures easily and is very brittle

LUSTER: resinous, like thin fibers

STREAK: white

HARDNESS: 1.5–2.5

OTHER DISTINCTIVE FEATURES: smells like rotten eggs, has a low density compared to other minerals

MATCH HEADS ARE MADE WITH THE ELEMENT **SULFUR,** WHICH IS WHY THEY OFTEN SMELL LIKE **ROTTEN EGGS** WHEN **THEY BURN.**

Elemental sulfur is easy to identify because of its bright yellow color and the fact that it's often found coating rocks in areas near active volcanoes.

Native sulfur can sometimes shine like glass.

Bet You Didn't Know!

Sulfur has a low melting point and is a poor conductor of heat. It will burn easily, producing a blue flame. Even the heat from your hand will cause it to change, and you can hear cracking sounds if you hold it up to your ear! The cracking happens because the outside of the mineral expands when it gets warm, but the inside stays the same size.

Native gold often forms flakey deposits that look a little like leaves!

Bet You Didn't Know!

Because pure gold is so soft, most gold jewelry is a mix of gold with other metals such as silver (pp. 146–147) to make an alloy. An alloy is a metal made from two or more metals mixed together. In many cases, an alloy is much harder and wears better than the pure metals alone. The gold content in a piece of jewelry is measured in karats (not to be confused with the carats used to measure the size of diamonds, or carrots used as a treat to feed your pet rabbit). A ring that is pure gold is 24 karats, while a 12-karat gold ring is half gold and half some other metal.

NATIVE GOLD

What's the first thing that you think of when you hear the world "gold"? Maybe it's a shiny necklace or a beautiful ring. How about a stack of gold coins or even some buried treasure bringing you wealth beyond your wildest dreams! Because of its value, gold is one of the most sought-after and expensive native elements on the planet.

Gold has a few interesting properties that set it apart from other metals and make it ideal for use in jewelry and decorative artwork. First, gold is shiny and, most important, stays shiny. The reason? Unlike other metals, it doesn't react with oxygen and water, so it does not tarnish or rust. This is very important if you want your "bling" to stay looking like bling! Second, not only is gold softer than most metals, it's also both malleable and ductile. This means it can be easily shaped by hammering and can also be drawn out into thin strands or wires. Finally, when gold is found in nature, it is often in a pure form, so it does not require a great deal of processing to remove unwanted materials.

Even though gold is a fairly rare find, it can be unearthed in many different locations around the world. Gold forms in veins in igneous rocks and is frequently found with quartz (pp. 206–207) and the minerals pyrite (pp. 160–161) and calcite (pp. 190–191). Most often, when gold is discovered, it's in the form of irregularly shaped nuggets or as a scaly covering on another rock. Gold can form well-developed crystals that look like double pyramids, but these are very unusual. One of the other properties that sets gold apart from other minerals is its high specific gravity, the measure of how heavy a mineral feels. Most minerals have a specific gravity close to 5, but for gold, it can be greater than 15. That makes it a real heavyweight among minerals!

FACTS

CLASS: native elements

MAIN COLORS: golden yellow, brassy yellow

CLEAVAGE: none

LUSTER: metallic

STREAK: gold

HARDNESS: 2.5–3

OTHER DISTINCTIVE FEATURES: very malleable, bends easily, extremely dense

GOLD IS SO MALLEABLE THAT IT CAN BE HAMMERED INTO SHEETS THAT ARE 0.1 MICRON THICK, OR ABOUT ONE-THOUSANDTH THE THICKNESS OF A SHEET OF STANDARD COPIER PAPER!

THE RUSH IS ON!

When it comes to native elements, the one that sets the "gold standard" by which all other metals are judged is, you guessed it ... gold! People have been fascinated with this shiny yellow metal since before recorded history, and many have gone to great lengths to find it. Even though gold can be found in many types of igneous rocks, it is present only in small amounts. This means that it may cost more money to get the gold out of the rock than the gold is worth. Sometimes hot, mineral-rich fluids rising through the rock can create deposits called veins that have a large amount of gold in them. When gold veins are discovered, the rock is chopped out and crushed, and a bunch of different chemical processes are used to separate the gold from the other minerals.

A much simpler method of finding gold is to have Mother Nature do the hard work of separating the gold from the rock and depositing it for you. Gold is very resistant to weathering, so when rocks that have gold in them wear away, the gold is often left behind. Rain and melting snow then wash the gold into streams. But because gold is so much heavier than other rocks and minerals, it is often left behind in small pockets at the bottom after the other rocks have been carried away by the water. These are called placer deposits, and they are usually found downstream from igneous rocks that have gold in them.

One of the ways people look for gold is by panning for it. They usually start by heading to a stream near where gold has already been found and locating a quiet spot where sediment has built up on the bottom. Using a large metal pan that looks like a big pie plate, they scoop up some of the sediment along with some water and gently slosh the pan around so some of the mud comes out. Because gold is so dense,

MOST **CHUNKS** OF **GOLD** TEND TO BE **PRETTY SMALL,** BUT ONE NUGGET CALLED **"THE MONUMENTAL,"** FOUND IN **CALIFORNIA IN 1869,** WEIGHED IN AT **106 POUNDS** (48 KG)— ABOUT AS MUCH AS A **10-YEAR-OLD KID!**

Bet You Didn't Know!

The gold bars that are stacked up in bank vaults in movies and television shows weigh a lot more than most people think. Known as "Good Delivery" bars, these standard units are officially set at 400 troy ounces, meaning that each one weighs 27.4 pounds (12.4 kg), heavier than a sack of potatoes!

One of the easiest ways to find gold is to go panning for it in a stream.

THERE'S **GOLD** IN THEM **THAR HILLS!**

Every so often, someone will find a large deposit of gold in a new area. Once the word gets out about the discovery, people come from all over to strike it rich, creating what is known as a gold rush. Two of the most famous gold rushes in the United States were the Klondike gold rush in Alaska and Canada during the late 1890s and the even more famous California gold rush that started in 1848.

The California gold rush started when some gold flakes were discovered in a place called Sutter's Mill, about 50 miles (80 km) from Sacramento. Even though John Sutter, who owned the property, tried to keep it a secret, word spread like wildfire, and by 1853 approximately 250,000 people had come to California seeking their fortune. Unfortunately, most of these people never struck it rich, and the people who made the most money were the folks who sold supplies and services to the gold seekers.

any pieces in the sediment will sink to the bottom of the pan. After repeating the process a few more times they check to see if there is anything shiny in the bottom of the pan. Gold flakes in sediment tend to be really tiny, so this process takes time. But if you can get enough of them, they'll be worth their weight in gold!

GOLD MINERS IN CALIFORNIA, 1850

145

NATIVE SILVER

When you hear stories of people searching for secret treasures, they almost always speak about finding gold AND silver. While it may not be quite as valuable as gold, silver is also a precious metal used to make jewelry and coins. It has many of the same properties as its brassy yellow cousin.

Like gold (pp. 142–143), native silver forms as a result of hot fluids moving up through Earth, often forming veins with quartz (pp. 206–207) in a variety of igneous rocks. While it can form crystals that look like cubes, eight-sided octahedrons, and 12-sided dodecahedrons, these tend to be rare. Most often native silver forms sheets and irregularly shaped masses in mafic igneous rocks along with copper (pp. 150–151). Even though it is not as dense as gold, silver still has a high specific gravity, so it feels heavy. It also is very flexible and can be hammered into sheets and stretched into wires, which is why it works well in making jewelry.

Unlike gold, silver reacts with other elements and will easily tarnish to a dark gray or even black when it is exposed to the air for a long time. While native gold rarely forms natural compounds with other elements, silver does form natural mixtures with other metals called alloys and can be found with copper, platinum (p. 148), antimony (p. 149), and bismuth (p. 149). Much of the silver that is mined in the world comes from other minerals that contain silver and have weathered and been changed over time.

Silver has many uses in industry, and because it is a very good conductor of electricity, it is often found in electronic components. Some folks have utensils like forks and knives made out of pure silver, but these objects are usually very expensive. Many people will opt for silver-plated eating utensils instead, which are made from another metal with a thin coating of silver on top. These are much less expensive and still look almost like they're pure native silver.

FACTS

CLASS: native elements

MAIN COLORS: silver white but may tarnish to dark gray

CLEAVAGE: none but will fracture with rough edges

LUSTER: metallic

STREAK: silver white to light gray

HARDNESS: 2.5–3

OTHER DISTINCTIVE FEATURES: feels heavy, with a specific gravity of 10+, malleable

DENTISTS USED TO **FILL CAVITIES** WITH A MATERIAL CALLED **AMALGAM**, WHICH IS AN ALLOY OF **MERCURY, SILVER, COPPER, AND TIN.** THESE DAYS, MOST DENTISTS PREFER TO USE A **POLYMER MATERIAL** THAT COSTS LESS AND **MATCHES TEETH** BETTER.

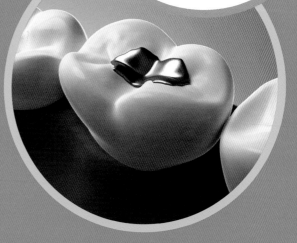

Native silver often will form as a crust on top of intrusive igneous rocks.

Native silver will often appear like little branches.

Bet You Didn't Know!

Before 1965, half dollars, quarters, and dimes in the United States were made of 90 percent silver and 10 percent copper. But because of the rising price of silver, in 1965 the U.S. Mint changed the composition of these coins to a combination of copper and nickel; by 1971, none of the coins being made contained silver. If you have one of these pre-1965 coins, make sure to save it because the silver in it makes it worth much more than its face value!

NATIVE AND RARE

Geologists have identified thousands of different minerals, each with its own unique structure and chemical composition. Most of these minerals are chemical compounds, meaning they are made from two or more different chemical elements joined together. Only about a dozen minerals are commonly found as individual native elements. Some, like gold and silver, are famous, but here are a few that you might not have heard of.

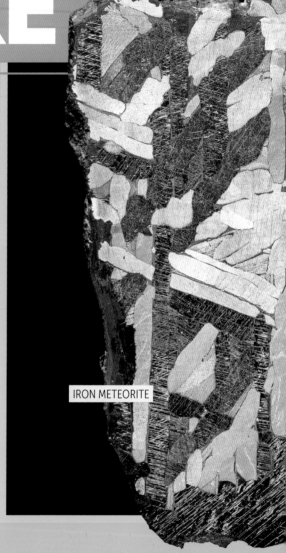

IRON METEORITE

IRON

You're probably wondering why iron is included in a list of uncommon native elements. After all, iron is a very common element found in many types of rocks. This is true, but almost all of that iron is found combined with other elements to make complex minerals. Only rarely does iron appear as a native element and, when it does, it is usually mixed with at least a small amount of nickel. There are two main sources of native iron. It can sometimes be found in basalt flows where it has become concentrated in small blebs (that is a technical term for a small bubble of one mineral inside another rock). The other common place to find native iron is in iron meteorites from space! Before people discovered how to get iron out of Earth rocks, they used this extraterrestrial iron to make tools and weapons. Iron is quite hard and, before it is exposed to the air, has a dark gray to black color. It will quickly get a reddish brown coating when exposed to the air as it begins to rust.

PLATINUM

Platinum, like gold and silver, is a precious metal that is used in making jewelry. This silvery metal usually forms in ultramafic igneous rocks along with olivine (pp. 234–235), pyroxene (pp. 224–225), and magnetite (pp. 172–173). Platinum has a specific gravity that can often reach 20, so it feels heavier than most pieces of gold. And with a hardness of 4 to 4.5, it is harder than most metals. Most platinum is found as sedimentary placer deposits along with gold (pp. 142–143) after it has weathered out of the original igneous rock that it formed in. The name platinum, which the metal was given because of its color, comes from the Spanish word *plata,* meaning "silver."

GRAPHITE

Graphite is the only native element on this page that is not a type of metal. That's because it is made of pure carbon. Found in a variety of metamorphic rocks including schist (p. 107) and gneiss (p. 106), graphite forms when sediments rich in carbon are heated or react with hot fluids that flow through them. Graphite can form six-sided crystals or large masses and, even though it is not a metal, it can have a metallic luster. It is usually jet black to gray in color and is super soft—so soft that it feels greasy and actually rubs off on your finger. Graphite powder is used as a "dry" lubricant to keep machine parts from sticking together, but the most common use is to make the black filling inside pencils.

NATIVE IRON

ANTIMONY

This is not a common metal, but it has been used for thousands of years to make alloys with other metals such as lead and tin. Antimony is found in almost 100 different minerals, often extracted from the mineral stibnite (p. 164). But it has also been found as a native element, usually with silver (pp. 146–147) and arsenic (p. 156). Like silver, it has a metallic luster and usually appears bluish white. It can form cubic crystals, but most often it just appears as small grains or a massive chunk. Though it is harder than gold or silver, antimony is fairly brittle, so it can easily crumble.

BISMUTH

Even though it is not a common mineral, people have known about bismuth and used it for more than 500 years. Often bismuth is found with quartz (pp. 206–207), gold (pp. 142–143), and cassiterite (p. 174). Bismuth can also form in pegmatite veins with quartz and in areas where hot fluids from deep in the Earth have replaced other minerals. Bismuth is soft and brittle, and when it forms crystals they are in parallel strips. Most often it has a metallic luster and appears silver-white with a reddish tinge. One interesting use of bismuth salts in the past was to make special drinks for people who had to have x-rays of their stomachs and intestines. Bismuth salts are also used in medicines that treat stomachaches.

Looking like the branches of a tiny tree, native copper is often found in mafic igneous rocks.

Bet You Didn't Know!

Copper tarnishes quickly when it is exposed to air, but instead of turning rusty red like iron or black like silver, it first turns a dull brown. Then, after it has been exposed to the air for a long enough time, the copper begins to turn green. That's what happened to the Statue of Liberty, designed by French sculptor Frédéric-Auguste Bartholdi and given to the United States as a gift from the people of France. Originally a shiny copper, the statue's classic green look is the result of its exposure to the salty air of New York Harbor for more than 100 years!

NATIVE COPPER

Even though it may not be as precious as gold, silver, or platinum, copper is one of the most important metals. People have been using it in its native form for thousands of years. Unlike other native metals, native copper is quite common and is found in many locations in the world. Most often it is found in lava flows of basalt (p. 90) as a secondary mineral, meaning that it didn't form when the rock originally hardened. Instead, chemically active fluids flowing through the rock reacted with iron-rich minerals in the basalt to produce pure copper, which was then deposited inside the rock.

While native copper can form crystals that look like cubes or soccer balls, it is most often found as scaly coverings or plates on other rocks. In some cases, the copper will form a fine branching pattern that looks like a tiny tree. Like native gold (pp. 142–143) and silver (pp. 146–147), copper is extremely malleable and can be easily shaped by hammering. It is also very ductile, which means that it can be drawn out to form thin strands or wires. Finally, copper doesn't bend quite as easily as either gold or silver, which makes it useful as a cutting tool. Archaeologists believe that copper was the first metal to be used for tools and weapons. In ancient Egypt, it was used to make knives and even saw blades that could cut through limestone (pp. 120–121). Later, people learned that they could mix copper with zinc to make brass, and with tin and sometimes another metal such as zinc to make bronze. These materials were even harder and more durable.

Today, copper is used in hundreds of different ways. Because it is a good conductor of both heat and electricity, it is used to make pots and pans for cooking and wires for electronic devices. Copper is also used to make pipes for water, and it is even used for crafting pieces of art. Even if copper isn't as precious as gold, it is much more useful and important in our daily lives.

FACTS

CLASS: native elements

MAIN COLORS: copper red on fresh surfaces, tarnishes to dull brown often with a greenish tint

CLEAVAGE: none; fractures with rough edges

LUSTER: metallic

STREAK: reddish

HARDNESS: 2.5–3

OTHER DISTINCTIVE FEATURES: very malleable and can be drawn out into thin strands

ANCIENT CYPRUS COPPER

THE **WORD "COPPER"** COMES FROM **KYPRIOS,** THE **GREEK NAME** FOR THE **ISLAND OF CYPRUS,** WHERE COPPER WAS **MINED** AS EARLY AS 4000 B.C.

DIAMOND

Diamonds are the stuff of legend. They are by far the most famous gemstone in the world and, when they are cut just right, can bounce and bend light like no other. People cherish them and give them as special gifts. Famous diamonds can be found on display in museums around the world.

Diamonds are made of pure carbon. Most form deep under Earth's surface, where high temperatures and tremendous pressure force the carbon atoms into a supercompact structure. This unique structure makes diamonds really hard. But diamonds also have zones of weakness in them called cleavage planes. These zones allow a skilled lapidary (a person who cuts and polishes gemstones) to cut and shape them into beautiful gems.

Diamonds form crystals that are shaped like cubes, eight-sided pyramids called octahedrons, and even 12-sided dodecahedrons, which look like little soccer balls. While the most common gem-quality diamonds are either clear or pale yellow, they can come in a variety of colors including pale red, orange, blue, green, and brown. One rare variety of diamond called carbonado is black and opaque, meaning light does not pass through it.

While gem-quality diamonds are special, this magnificent mineral also has a practical side. Because it is the hardest mineral, diamonds are also used for lots of industrial purposes. As you might expect, industrial diamonds do not have the same "look" as gem-quality diamonds. Diamonds that are badly colored and have a rounded or irregular appearance with rough edges are called bort, and it's these that usually wind up as industrial diamonds.

Because diamonds are so hard, their most common industrial use is for cutting and polishing. Diamond-edged saw blades are used to cut stone and glass, and diamond dust is used for polishing other stones, including other diamonds! Diamond-tipped drill bits are used when drilling for oil and for cutting steel.

Yes, diamonds are truly spectacular. Even the name is special. The word "diamond" comes from the Greek word *adamas,* which means "invincible"—a proper name for the hardest of all minerals!

FACTS

CLASS: native elements

MAIN COLORS: clear and pale yellow are most common, but can have a red, blue, or green tint

CLEAVAGE: perfect in one direction

LUSTER: brilliant, glassy

STREAK: none

HARDNESS: 10

OTHER DISTINCTIVE FEATURES: crystals are often shaped like double pyramids or octahedrons

CARBONADO, WHICH IS SOMETIMES CALLED A **BLACK DIAMOND,** IS RARE. SOME GEOLOGISTS THINK THEY FORM FROM THE **HEAT AND PRESSURE OF METEORITE IMPACTS.** ANOTHER THEORY SAYS THAT THEY MAY HAVE FORMED **INSIDE OF A SUPERNOVA** AND WERE CARRIED TO EARTH **BY ASTEROIDS!**

Not all diamonds are "gem quality." In fact, most diamonds are used for industrial purposes instead of jewelry.

No other gemstone sparkles like a diamond that's been cut and polished.

Bet You Didn't Know!

Diamond and graphite are both native elements made from pure carbon, but they are nothing alike. Graphite is so soft that you can break it apart with your fingers, while diamond is the hardest mineral known. The reason for the difference is the way the carbon atoms are connected together. In a diamond, the atoms form a tight network, with each carbon joined to four other carbon atoms by strong bonds. In graphite, the atoms form sheets that are only weakly bonded together.

DIGGING UP DIAMONDS

PACIFIC OCEAN

Have you ever dreamed of digging up your own diamonds? Plenty of people have. But it turns out that, like many natural resources, gem-quality diamonds are not evenly spread around the world. Some countries are chock-full of diamonds, and others have none at all. This makes a great deal of sense when you consider the way diamonds form.

Diamonds can form only under high temperatures and enormous pressures not typically found in Earth's crust. Many geologists think that diamonds actually form in the mantle more than 75 miles (120 km) below the surface, where the carbon atoms can get squeezed into a supertight mineral structure. From there, diamonds are carried to the surface by magma that flows up through narrow passageways in the crust called kimberlite pipes. These are named for Kimberley, South Africa, where one of the deepest diamond mines was dug. Most often, diamonds found surrounding kimberlite pipes are in an ultramafic igneous rock called peridotite.

Another place where diamonds are frequently found is in layers of gravel that were deposited by streams that flow down from areas where peridotite rocks are located. It turns out that, because diamonds are fairly dense and are super hard, they are left behind when the peridotite rock weathers away. They are then moved downstream by flowing water and collect at the bottom of streams in the same way that gold (pp. 142–143) forms placer deposits.

Diamond-producing countries form two separate belts around the world. In the Northern Hemisphere, both Russia and Canada produce a large amount of gem-quality diamonds. In the

DIAMONDS WERE FIRST **ACTIVELY MINED** IN THE **FOURTH CENTURY** B.C. IN **INDIA,** WHICH **STILL PRODUCES** DIAMONDS **TODAY.**

Bet You Didn't Know!

Even though the United States is not a major diamond-producing nation, it does have a few locations where diamonds can be found. The most famous is called Crater of Diamonds State Park located near Murfreesboro, Arkansas. For a fee, people can try their luck at digging up their own diamonds, and the cool part is that you get to keep what you find!

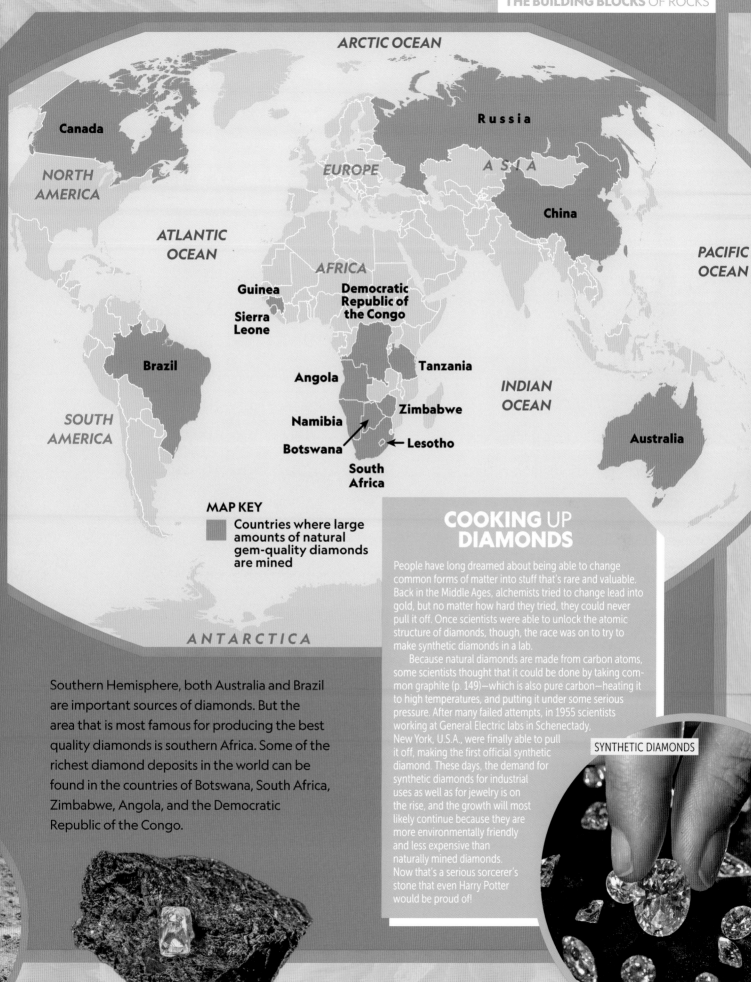

ARCTIC OCEAN

Canada

NORTH AMERICA

ATLANTIC OCEAN

SOUTH AMERICA

Brazil

EUROPE

Russia

ASIA

China

AFRICA

Guinea

Sierra Leone

Democratic Republic of the Congo

Angola

Namibia

Botswana

Tanzania

Zimbabwe

← Lesotho

South Africa

INDIAN OCEAN

PACIFIC OCEAN

Australia

ANTARCTICA

MAP KEY

Countries where large amounts of natural gem-quality diamonds are mined

Southern Hemisphere, both Australia and Brazil are important sources of diamonds. But the area that is most famous for producing the best quality diamonds is southern Africa. Some of the richest diamond deposits in the world can be found in the countries of Botswana, South Africa, Zimbabwe, Angola, and the Democratic Republic of the Congo.

COOKING UP DIAMONDS

People have long dreamed about being able to change common forms of matter into stuff that's rare and valuable. Back in the Middle Ages, alchemists tried to change lead into gold, but no matter how hard they tried, they could never pull it off. Once scientists were able to unlock the atomic structure of diamonds, though, the race was on to try to make synthetic diamonds in a lab.

Because natural diamonds are made from carbon atoms, some scientists thought that it could be done by taking common graphite (p. 149)—which is also pure carbon—heating it to high temperatures, and putting it under some serious pressure. After many failed attempts, in 1955 scientists working at General Electric labs in Schenectady, New York, U.S.A., were finally able to pull it off, making the first official synthetic diamond. These days, the demand for synthetic diamonds for industrial uses as well as for jewelry is on the rise, and the growth will most likely continue because they are more environmentally friendly and less expensive than naturally mined diamonds. Now that's a serious sorcerer's stone that even Harry Potter would be proud of!

SYNTHETIC DIAMONDS

DANGER! TOXIC ROCKS AHEAD

SAFETY TIP!

BEWARE! The rocks on these pages can be poisonous to touch, smell, and taste. You should not handle or interact with them in any way.

ALERT AN ADULT if you think you've found one of these rocks. Find an adult and tell them what you think you've found. Remember that you should not handle or interact with these rocks in any way.

Most minerals are totally safe to handle and fun to examine and study. Some, like halite (pp. 186–187), are even safe to eat (and taste great on french fries). But just because something is from the natural world, that doesn't mean it's safe. Some minerals can be dangerous if handled improperly—and a few can actually kill you if you accidentally ingest them. Here's a rundown of some of the potentially toxic rocks that you should avoid if you ever encounter them.

ORPIMENT

With its yellow-orange color and honey-like luster, orpiment (which contains the element arsenic) looks harmless but is actually quite dangerous. Because it is quite soft and easily breaks down into a powder, it can irritate the skin. And breathing in the dust can make you very ill. Orpiment rarely forms crystals and can be found near hot springs. It also forms when other arsenic minerals such as realgar break down. Even though it smells a little like garlic, you should never bring it near your nose or mouth.

ARSENIC

Arsenic is a native element that has been a known poison since ancient times. Compounds made from arsenic are used for killing weeds, bacteria, and even rats. The biggest dangers from handling native arsenic come from inhaling the dust or accidentally eating it, which is why you should never, ever taste rocks. Finding native arsenic is rare, because the element usually forms compounds with other elements. It is most often discovered in igneous rocks near veins with silver (pp. 146–147) and antimony (p. 149) and is often a dark gray color with a metallic or earthy luster, depending on how weathered it is.

ARSENOPYRITE

The name alone should serve as a warning for this mineral! Arsenopyrite is the most common ore for the poisonous element arsenic. It usually has a silver-white color with a metallic luster and black streak and is often found with galena (pp. 158–159) and pyrite (pp. 160–161), forming when hot mineral-rich water flows through cracks in a rock. Crystals of arsenopyrite can be long and shaped like prisms. They are nice to look at but should not be handled and should never be brought near your nose or mouth.

Orpiment and realgar are often found together in the same rock.

CINNABAR

People are often attracted to the mineral cinnabar because of its bright red color and the fact that it can form nice six-sided crystals. When it comes to minerals though, looks can be deceiving: It's best to treat cinnabar as a "pretty poison." Cinnabar is one of the main sources of the metal mercury and is often found with pyrite (pp. 160–161), marcasite (pp. 162–163), and stibnite (p. 164). Mercury was once considered to be safe but is now known to be very dangerous, especially if you breathe in vapors and dust particles that are coated with it. The name "cinnabar" comes from the Arabic term *zinjafr,* which means "dragon's blood"—probably due to its red color.

REALGAR

With its bright red and orange color, realgar really stands out when compared to most other minerals. But these colors should serve as a warning sign to rock hounds: Realgar is an important ore of the element arsenic. Like native arsenic, realgar should be handled with great care and you should never bring it near your mouth or nose. Realgar can form crystals but it is sometimes found as a crust covering rocks near hot springs and volcanoes. It often breaks down and forms a powder when it is exposed to light.

Galena is easily identified by its dark gray color, cube-shaped crystals, and shiny metallic luster.

Bet You Didn't Know!

Galena looks very similar in shape to the mineral halite (pp. 186–187), which is commonly called table salt. They both form cubic crystals and break with square edges. It turns out that they have the exact same atomic structure, only with different chemical elements. Because galena is made from lead instead of sodium, it is much heavier than salt and is not transparent. That makes it easy to tell the two minerals apart.

GALENA

If you're into heavy metal, then we've got a mineral for you. (We're not suddenly talking about rock music here—although "rock" music is what a geologist would probably listen to!) When we say heavy metal, we're literally talking about lead, which is a superdense heavy metal that galena is made from.

Galena is in the sulfide class of minerals, which all include the element sulfur in their chemical composition. Besides having a high specific gravity, making it feel heavier than most minerals, galena has a few other properties that make it easy to recognize.

First, it's a dark gray color and often has a shiny metallic luster, especially on recently broken surfaces that have only been exposed to the air for a short amount of time. Galena is also quite soft, so it can easily be scratched with the point of a steel nail. Finally, galena crystals often are shaped like cubes or a combination of cubes and eight-sided octahedrons.

Galena is found in a variety of rocks, but it is most common in igneous veins along with other sulfide minerals including pyrite (pp. 160–161) and marcasite (pp. 162–163). It can also be found with both quartz (pp. 206–207) and calcite (pp. 190–191). In some cases, galena can be found in sedimentary rocks that have undergone contact metamorphism, when magma that created pegmatite flowed through them.

Galena's main use is as an ore of lead. Lead is no longer used as an additive in paints and gasoline or to make pipes for drinking water, but it is still found in a variety of products. It is still used to make waste pipes and solder, an item used to hold components in place in electronic devices.

FACTS

CLASS: sulfide minerals

MAIN COLORS: dark gray

CLEAVAGE: perfect in three directions at right angles to each other

LUSTER: metallic, shiny

STREAK: dark gray

HARDNESS: 2.5

OTHER DISTINCTIVE FEATURES: smells a little like rotten eggs, very heavy compared to other minerals

GALENA IS OFTEN **FOUND** WITH **DEPOSITS** OF **COPPER, ZINC,** AND **ANTIMONY** AND MAY ALSO CONTAIN **SOME SILVER.**

PYRITE

Imagine that it's 1849 and you just arrived in California, U.S.A., to seek your fortune as a gold prospector. You get off the stagecoach and are approached by a good-natured fellow who explains that he has to get back East and has no money, but he does have some gold that he would be willing to sell you cheap. He takes out a few chunks of a gold-colored rock and, thinking that you just made the deal of a lifetime, you pay him and take the "gold." You then hurry to the local claims office to cash in your bonanza, only to find out that you just spent all your money on a few ounces of worthless iron pyrite.

During the gold rush, some people were tricked into thinking that pyrite was actual gold, which is why it is commonly called fool's gold. The truth is, if you know a little about mineral properties, it's almost impossible to confuse gold for pyrite. First, gold is much softer than pyrite, which contains iron and can actually produce sparks when you hit it with a piece of flint. Second, gold is three times denser than pyrite, which has a specific gravity of only 5. Pyrite feels much lighter compared with a piece of gold that is the same size. Finally, when you rub gold on a streak plate it leaves a gold mark. Pyrite has a black or greenish black streak.

Pyrite is actually the most common type of sulfide mineral. It can be found in many rock types, including igneous pegmatite veins along with quartz (pp. 206–207) and microcline feldspar (pp. 212–213). Pyrite is found in rocks that have undergone contact metamorphism or that have been altered by hot, mineral-rich water moving through them. Pyrite is also found as a secondary mineral in many sedimentary rocks, forming as the original minerals go through chemical changes after the rock has formed.

Unlike gold, pyrite usually forms spectacular crystals that are often shaped like perfect cubes or octahedron. Unfortunately, pyrite is much less stable than gold and will tarnish quickly. It will even turn into other minerals such as limonite (pp. 176–177) when exposed to the air. The bottom line is that if some of the unfortunate prospectors had known what you know now, they never would have been fooled into paying for worthless "fool's gold"!

PYRITE GETS ITS NAME FROM THE **GREEK TERM PYR,** MEANING **"FIRE,"** BECAUSE THE **IRON** IN PYRITE **WILL MAKE SPARKS** WHEN IT IS HIT AGAINST A **HARD ROCK** SUCH AS **FLINT** (PP. 130–131).

Because of its golden color, pyrite is sometimes confused with gold.

Bet You Didn't Know!

Can you imagine finding fossils made of gold? Well that's what people first thought they had found when they came across pyritized seashells. Pyrite forms as a secondary mineral in sedimentary rocks and can sometimes replace the original minerals by a process called permineralization, creating a gold-colored fossil.

Marcasite has a similar color to its chemical twin pyrite, but its crystals have a totally different shape.

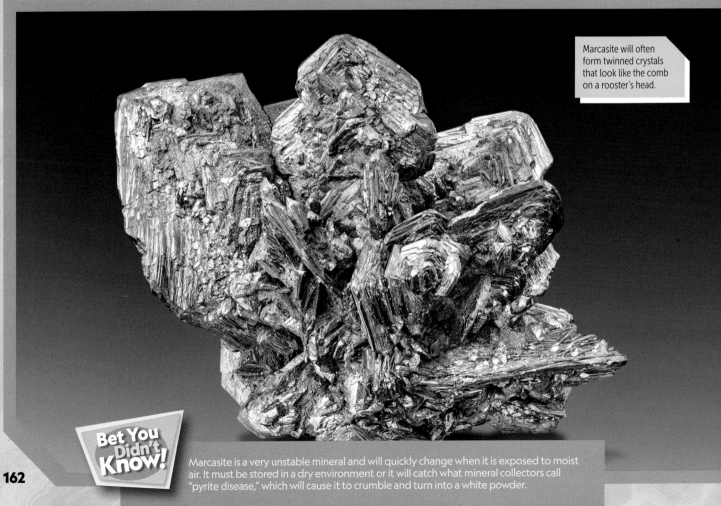

Marcasite will often form twinned crystals that look like the comb on a rooster's head.

Bet You Didn't Know!

162

Marcasite is a very unstable mineral and will quickly change when it is exposed to moist air. It must be stored in a dry environment or it will catch what mineral collectors call "pyrite disease," which will cause it to crumble and turn into a white powder.

MARCASITE

Marcasite and pyrite (pp. 160–161) are similar in that they both are sulfide minerals with the exact same chemical composition. They also have similar colors, although pyrite tends to be golden yellow with a black or greenish black streak, while marcasite, which has a dark gray streak, is pale yellow and even silvery on freshly broken surfaces that have not been exposed to air. Because of this coloring, marcasite is sometimes called white pyrite, but the two have some major differences that set them apart.

First, the two minerals have totally different crystal structures. Pyrite forms crystals that look like cubes, but marcasite crystals tend to be long, curved, and pointed like stalks of wheat. Also, when marcasite crystals grow, they often form twins, which means that two crystals grow from the same place in opposite directions. When this happens, the crystals stack up next to each other and look almost like the comb on a rooster's head.

Pyrite and marcasite also tend to form in different environments. While pyrite and marcasite can both form as a result of hot, mineral-rich waters pushing their way up toward the surface, marcasite often forms as a result of naturally acidic water flowing down from the surface through beds of shale (p. 127) and limestone (pp. 120–121), replacing some of the original minerals. Marcasite can frequently be found as lumpy nodules in clay and chalk (pp. 122–123). Like pyrite, marcasite will often replace the calcite minerals in shells, producing a fossil that has a golden color.

In the past, marcasite was used as a source of sulfur, but these days it has little value as an ore. The name "marcasite" is also used to describe bracelets and other types of jewelry, but because it easily tarnishes, most of these pieces are actually made from pyrite, which is much more durable.

FACTS

CLASS: sulfide minerals

MAIN COLORS: pale bronze-yellow

CLEAVAGE: two directions but not dominant; brittle with uneven fracture

LUSTER: metallic

STREAK: dark gray

HARDNESS: 6–6.5

OTHER DISTINCTIVE FEATURES: tarnishes to a dark yellowish brown

WARNING!

WHEN **MARCASITE** IS **EXPOSED TO MOISTURE** AND BREAKS DOWN IT CAN FORM **SULFURIC ACID,** WHICH IS USED TO MAKE **CAR BATTERIES** AND CAN **BURN YOUR SKIN.**

SUPER SULFIDES!

Sulfide minerals are simple chemical compounds that are a combination of sulfur with another element that is often a metal. Many sulfide minerals are important ores for the metals they contain and are considered to be valuable natural resources. Some also have really cool-looking crystals. Here's a look at five super sulfide minerals and what they are used for.

STIBNITE

The main ore for the metal antimony, stibnite can be found in limestone (pp. 120–121), gneiss (p. 106), and granite (pp. 78–79) rocks. It is used for making a wide range of products including matches, fireworks, and rubber; in the past, people used it in makeup to make their eyes look bigger. Usually dark gray to silver gray in color, stibnite has a shiny metallic luster on surfaces that have been recently broken and have not been exposed to the air. This mineral is fairly soft, so it can be scratched with a penny. Stibnite usually forms long, thin pointed crystals that are often bent or twisted—sometimes resembling blades of grass growing out of a single clump.

CHALCOPYRITE

Chalcopyrite is the most common mineral containing the element copper. But, like the mineral pyrite (pp. 160–161), it also has iron in it (hence its name). It is frequently found in veins deposited by hot, mineral-rich water along with pyrite and gold (pp. 142–143), forming massive coatings on the surrounding rock. Like pyrite, it has a shiny metallic luster and a brassy yellow color on freshly broken surfaces, but when it is exposed to the air, it tarnishes to have an iridescent finish that can be shades of blue, purple, and green. Chalcopyrite does not always form crystals, but when it does the crystals are usually shaped like pyramids. Because there is so much of it in the world, chalcopyrite is an important ore of copper.

MOLYBDENITE

Molybdenite forms when the metal molybdenum joins with sulfur. It is often found in veins of granite pegmatite or in "dirty limestones" (pp. 120–121) that have undergone high-temperature contact metamorphism. Molybdenite is the most important ore for the element molybdenum, which is used to strengthen steel and other metals. This shiny gray mineral is quite soft and almost feels like graphite (p. 149). However, it is easy to tell the two apart because molybdenite often forms six-sided crystals that almost look like a bunch of plates stacked on top of each other, while graphite's crystals are short and stocky.

SPHALERITE

Sphalerite is the main mineral ore of the element zinc, which is used to make batteries and to coat steel to keep it from rusting. It is usually found with galena (pp. 158–159) in many types of rocks, including igneous veins and rocks that have undergone contact metamorphism when hot, mineral-rich fluids moved through them. When it is pure, it looks white with a honey-like luster, but because it can also contain iron, it can be yellow, brown, or even black as the amount of iron in it increases. Sphalerite will often form complex crystal shapes that are combinations of cubes, pyramids, and 12-sided dodecahedrons. It is harder than many other sulfide minerals because the atoms join together with very strong bonds, creating a structure similar to that of diamond (pp. 152–153).

BORNITE

Like chalcopyrite, bornite is an ore for copper and forms when the elements copper, iron, and sulfur join together. It usually is found along with chalcopyrite, pyrite (pp. 160–161), and quartz (pp. 206–207) in veins deposited by hot, mineral-rich water flowing through rocks. Bornite rarely forms crystals. Instead it forms massive coatings on other rocks and, on freshly broken surfaces, has a copper-red color and a shiny metallic luster. When it is exposed to air, the surface quickly changes to become purple and blue, making it one of the most colorful minerals. That characteristic has earned it the name "peacock ore"!

CORUNDUM

When you hear someone describe the mineral corundum, you might think it's not all that exciting because in its pure form it is sort of a pale yellow or white. But if a small amount of the element chromium gets mixed in with the aluminum and oxygen that corundum is made from, the mineral takes on an entirely new look. That's when ordinary corundum turns into the gemstone ruby (pp. 168–169). If instead of chromium some iron and titanium get added into the mix, then the gemstone sapphire (pp. 168–169) is produced.

Actually, even plain corundum has a lot of interesting physical properties that set it apart from many minerals. Besides its gem-forming red and blue types, it can also come in brown, pink, and even green. Corundum is also extremely hard and serves as the index mineral for the number 9 on the Mohs scale. It also has a very high melting point and is quite dense. This combination of properties means that, like diamond, when corundum weathers out of rocks it often collects in placer deposits in streams and rivers.

Corundum forms in many types of rocks including felsic igneous rocks, especially pegmatites, and also in high-grade metamorphic rocks such as mica schist (p. 107) and gneiss (p. 106). It usually has six-sided crystals that form prisms or double-sided pyramids. Many of the crystals show lines called striations on their flat faces.

Because it is so hard, the main use of corundum is as an abrasive material used for polishing and grinding glass and metals. Often corundum is not pure and is found mixed with minerals such as hematite (p. 174), magnetite (pp. 172–173), and spinel (pp. 170–171). In this case, it's called emery, and it is used to make sandpaper and special polishing cloths. Before the invention of synthetic corundum, it was used for making emery boards, which people use to file their nails. When all things are considered, corundum is not so ordinary after all—and you could very well wind up wearing a piece of it as jewelry someday!

FACTS

CLASS: oxide minerals

MAIN COLORS: white, pink, gray, green, brown, black, red (ruby), blue (sapphire)

CLEAVAGE: none; uneven fracture, sometimes conchoidal

LUSTER: glassy

STREAK: white

HARDNESS: 9

OTHER DISTINCTIVE FEATURES: crystals often transparent or translucent

GEM-QUALITY CORUNDUM CRYSTALS ARE SOMETIMES USED IN WATCHES AND SCIENTIFIC INSTRUMENTS.

Corundum often forms six-sided crystals and comes in a wide range of colors, but it can also be colorless.

Bet You Didn't Know!

Compared with most minerals, corundum has a very high melting point: 3700°F (2040°C). As a result, it is used to make insulating bricks called refractories for the inside of kilns and furnaces and to make molds for hot molten metal in factories.

HARD-NOSED GEMSTONES

When it comes to precious gemstones, rubies and sapphires rank right near the top, along with diamonds (pp. 152–153), in popularity. Not only are all three spectacular to look at, but they are all also extremely hard minerals, so they are quite durable. To be called a true ruby, a piece of corundum must have a distinctly red color, and the most valuable are a deep blood-red. Rubies are found only in a few locations in the world, most often in Myanmar (Burma), Thailand, and Sri Lanka. Sometimes other red stones including garnet (pp. 242–243) and tourmaline (pp. 232–233) will be sold as rubies, but they don't have the same color or clarity as true rubies. One of the interesting properties of a true ruby is that when it is heated to a high temperature, it will turn green, but then turn back to red when it cools off.

The term "sapphire" is used to describe any colored gem-quality piece of corundum that is not red. Most people think of sapphires as being blue—which they are—but they can also be green, violet, gray, yellow, and even pink. Unlike rubies, which tend to be uniform in color, sapphires can be unevenly colored and will often have tiny imperfections in them that create some interesting effects in the light. One variety called alexandrite sapphire (which can also be made in a lab) actually changes color depending on the type of

THE **LARGEST BLUE SAPPHIRE** DISCOVERED TO DATE WAS FOUND IN A **MINE** IN **SRI LANKA.** IT WEIGHS IN AT A WHOPPING **1,404 CARATS,** WHICH IS ABOUT **10 OUNCES** (280 G)—AS MUCH AS A **LARGE POTATO!**

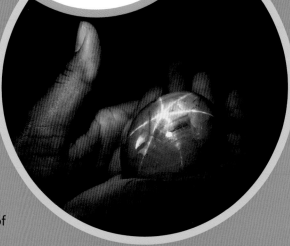

Bet You Didn't Know!

When Theodore Maiman created the first laser, he used a rod made of artificial ruby to produce the light. Using a device called a flashlamp, he was able to energize tiny particles called electrons in the chromium atoms of the ruby. These electrons then gave off pure red light that was concentrated by mirrors into a single beam and presto, the laser was born!

ALEXANDRITE SAPPHIRE

SAPPHIRES

Ruby and sapphire are popular gemstones that are actually colorful versions of the mineral corundum.

TRYING TO
MIMIC NATURE

Finding natural gem-quality rubies and sapphires is not easy. After all, one of the reasons that these minerals are considered gems is the fact that they are quite rare. To meet the increasing demand for these gemstones, a clever French chemist named Auguste Verneuil developed a method for creating synthetic rubies and sapphires way back in 1902.

Verneuil discovered that he could take superfine alumina powder (which contains both aluminum and oxygen) and heat it in an extremely hot flame until it melted. He would then allow the molten mass to drip down through a fine mesh and collect into a small mass called a boule, where it would cool and form a crystal with the exact same physical properties as the mineral corundum. Adding a bit of chromium or titanium to the mix would result in a ruby or sapphire that could be cut and polished just like a natural gem. Today, this "Verneuil process" is also used to make the minerals spinel (pp. 170–171) and rutile (p. 183).

light that it is viewed under. In daylight it is blue, but indoor lighting makes it appear violet or even red. "Star sapphires" get their name from tiny crystals of the mineral rutile (p. 183) that line up inside the sapphire, creating a white star in the middle of the gem.

As you might expect, because they form the same way, sapphires can be found in many of the same locations as rubies, but they also come from Australia, India, Russia, South Africa, and even the United States. Both rubies and sapphires are used to make rings and gemstones called cabochons, which are used to make pins and pendants. Any way you cut them, rubies and sapphires are a hard act to follow!

RUBY

The mineral spinel forms eight-sided crystals that often join together to form "twins."

Bet You Didn't Know!

When looking up information on the mineral spinel, you need to be careful because the name "spinel" has two different meanings. Not only is it used for the individual mineral described above, but it also can be used to describe a group of 10 related minerals, each of which has the same crystal structure as spinel, but with a different combination of metals in their chemical formula.

SPINEL

The mineral spinel contains the metals magnesium and aluminum. Like the other members of the spinel group of minerals, it is called an oxide because it also contains the element oxygen. Spinel is a very hard mineral, often hitting 8 on the Mohs scale. While it can appear as massive crusts, spinel will form nicely developed crystals, usually taking the shape of eight-sided octahedrons. Quite often, the crystals will form "twins" growing in two directions at once from a single starting point.

Spinel is a common mineral and can occur in a wide range of rock types. It's found in mafic and ultramafic igneous rocks including basalt (p. 90), peridotite, and kimberlites. Spinel also forms in marble (pp. 100–101) created from metamorphosed limestones (pp. 120–121), gneiss (p. 106), and schist (p. 107). It is also common to find spinel crystals mixed in with gravel and sand deposited by flowing water downstream from the igneous or metamorphic rocks they weathered out of.

Because of its glassy luster and transparent crystals, spinel is often used as an inexpensive gemstone. As with the mineral corundum (pp. 166–167), spinel can also be made artificially. Because there are so many possible combinations of metals that can be placed in the spinel structure, scientists have created more than 200 different synthetic spinel minerals.

FACTS

CLASS: oxide minerals

MAIN COLORS: lavender, brown, black, red, blue, green

CLEAVAGE: none; uneven fracture sometimes conchoidal

LUSTER: glassy

STREAK: white to grayish white

HARDNESS: 7.5–8

OTHER DISTINCTIVE FEATURES: crystals often transparent or translucent, forming twins

ONE DARK RED VARIETY OF SPINEL—CALLED **RUBY SPINEL**—IS OFTEN MISTAKEN FOR ACTUAL RUBIES. IN FACT, THE **TIMUR RUBY** THAT IS PART OF THE **CROWN JEWELS** OF THE UNITED KINGDOM IS ACTUALLY A **RUBY SPINEL.**

MAGNETITE

When magnetite was first discovered thousands of years ago, people probably believed these heavy black rocks—which are able to attract objects made of iron and steel—were magical. But, as you probably guessed from its name, magnetite is more magnet than magic.

Magnetite is an oxide mineral made from the elements iron and oxygen, and it is the iron that can make it behave like a magnet. It can form eight-sided octahedron crystals, but it usually is just unearthed in massive chunks. Magnetite is found in both igneous and metamorphic rocks that have been exposed to very high temperatures and have a lot of iron in them.

Not all pieces of magnetite will act like a magnet. In fact, most will only be attracted to a magnet. Those pieces that are strongly magnetized have a special name. They're called lodestones, which comes from two Old English terms that basically mean "stone to lead the way." It was first used for this purpose when people discovered that some chunks of magnetite could be turned into a simple magnetic compass to help them find north.

As you might expect, since it's loaded with iron, the main use of magnetite is as an iron ore. One fun thing that you can try is to collect your own magnetite sand. Next time you go to the beach, bring along a magnet and a plastic sandwich bag. Turn the bag inside out and put the magnet inside. Run the bag with the magnet through the sand and, if you're lucky, you'll have little pieces of black magnetite sticking to the magnet. When you've collected a bunch of magnetite sand, turn the bag right side out again and you'll have a bag of "magic minerals" to call your own!

FACTS

CLASS: oxide minerals

MAIN COLORS: black

CLEAVAGE: none; uneven fracture sometimes conchoidal

LUSTER: metallic or dull

STREAK: black

HARDNESS: 5.5–6.5

OTHER DISTINCTIVE FEATURES: strongly magnetic

YOU CAN MAKE A **SIMPLE COMPASS** BY TAKING A PIECE OF **LODESTONE** AND **TYING IT** TO A **STRING.** IF YOU LET IT HANG STILL, ONE END OF THE ROCK WILL **ALWAYS POINT** TO EARTH'S **MAGNETIC NORTH POLE.**

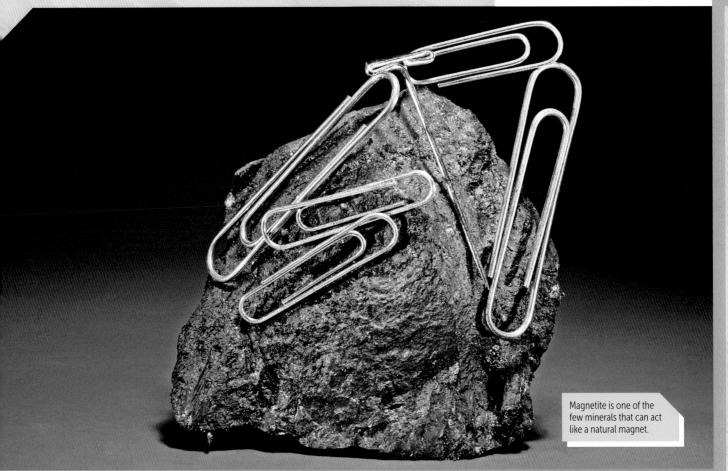

Magnetite is one of the few minerals that can act like a natural magnet.

Bet You Didn't Know!

Scientists still aren't sure where lodestone gets its magnetic powers from, but one theory has to do with lightning. Scientists have discovered that when lightning strikes the Earth, it can also produce a really powerful magnetic field that is strong enough to turn simple magnetite into lodestone.

173

METALMAKING MINERALS

Some oxide minerals such as corundum (pp. 166–167) and spinel (pp. 170–171) are valuable for producing beautiful gemstones, but this class of minerals has a practical side, too. Many of the most important metals that we depend on in our day-to-day lives—including iron, aluminum, titanium, tin, and chrome—come from oxide minerals. Here's a look at five of the top metalmaking minerals.

HEMATITE

Of all the oxide ore minerals, hematite is the hands-down winner when it comes to having different "looks." It is the most important source of iron, which can either be used directly as a metal or turned into steel. Hematite can be reddish brown, black, and dark gray, having either a dull or a metallic luster. What's interesting is that even the black varieties leave a reddish brown streak that is very helpful in identifying it. Hematite can produce a wide range of crystals, from thin plates and six-sided rhombohedrons that look a little like bent cubes, to botryoidal forms, which look like a bunch of grapes. Hematite is usually found in sedimentary rocks, where it forms from the breakdown of other iron-rich minerals. The red variety is sometimes ground up to give red paint its color.

CASSITERITE

The mineral cassiterite is also called tinstone. As you might have guessed, it is a major ore of tin, which is used in making metal alloys such as pewter and bronze. Tin is also used to coat other metals to keep them from rusting. Cassiterite forms long prisms or pyramid-shaped crystals with a greasy or dull to submetallic luster and comes in a variety of colors including brown, black, and (rarely) white. It is a hard mineral that is usually found in veins of granite pegmatite that have had hot, mineral-rich fluids running through them.

PLATY HEMATITE
CRYSTALS

BAUXITE

Technically speaking, bauxite is not a mineral. It is a rock that is a mixture of several different minerals that form when other minerals that are rich in the element aluminum change over time as they are exposed to air and water. Bauxite is the most important ore of the metal aluminum, which is used to make thousands of products that we use every day including foil to wrap foods, pots to cook foods, and lightweight beverage containers. Bauxite is usually found in tropical regions and is so soft that you can break it apart with your hands. It usually forms clumps or layers of rounded particles that can be white, gray, red, or yellow-orange in color.

ILMENITE

Ilmenite gets its name from the Ilmen Mountains in Russia, where it was first discovered. It is a major source of titanium, which—because it is both lightweight and super strong—is used in making aircraft. When ilmenite forms crystals, they can sometimes look like six-sided flat plates. But most of the time, ilmenite forms large, black masses that have a metallic luster. Ilmenite is usually found in mafic igneous rocks such as gabbro (pp. 82–83) and diorite (pp. 80–81) and is frequently mixed with the mineral magnetite (pp. 172–173).

CHROMITE

Chromite is an oxide that contains both iron and chromium and is a member of the spinel (pp. 170–171) group of minerals. It is the most important ore for the metal chromium, which is used to coat other metals to keep them from rusting. And it's what makes the bumpers of old-timey cars look so shiny! Because it is related to spinel, chromite can form eight-sided octahedron crystals, but this dark brown or black mineral is usually found as massive chunks in ultramafic igneous rocks such as peridotite or in a metamorphic rock called serpentinite. Sometimes, chromite can even be found as tiny crystals inside diamonds, but this does not happen often.

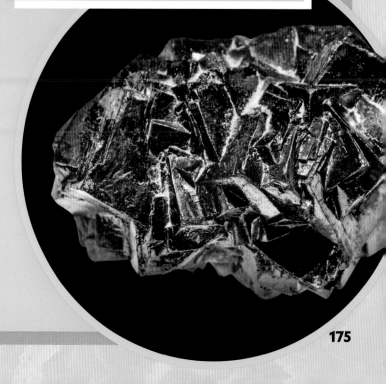

175

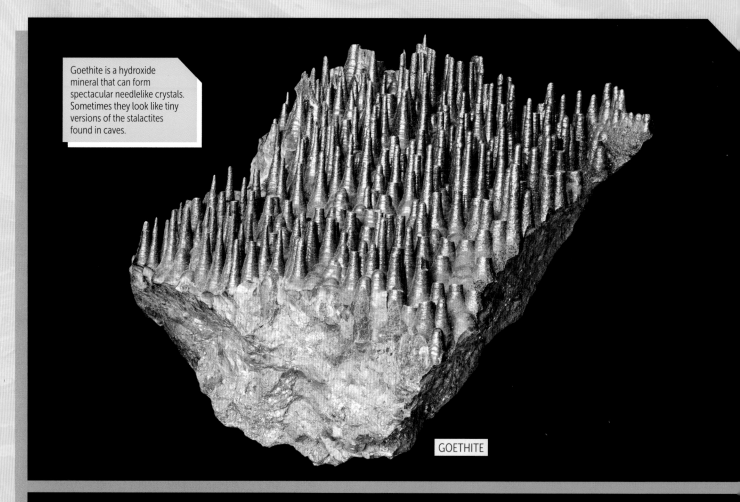

Goethite is a hydroxide mineral that can form spectacular needlelike crystals. Sometimes they look like tiny versions of the stalactites found in caves.

GOETHITE

LIMONITE

Goethite was named for the German mineralogist Johann Wolfgang von Goethe, who lived between 1749 and 1832. In addition to his work with minerals, Goethe was an expert in human anatomy and plants. His biggest claim to fame, though, was as a poet and playwright best known for *Faust*—a tragic play considered by many to be the greatest work in German literature.

LIMONITE/ GOETHITE

You may be wondering why one mineral has two names. Before the development of x-rays, limonite and goethite were thought to be different minerals. But, as frequently happens in science, advancements in technology gave us new information that caused us to make an adjustment in our thinking.

There were some good reasons why the two were thought to be different minerals, even though they have the same chemical composition. First, goethite usually appears dark brown, reddish brown, or yellowish brown, while limonite is mostly brownish yellow. Second, goethite is usually found in veins of iron-rich minerals such as pyrite (pp. 160–161) and magnetite (pp. 172–173) and forms as a result of weathering, while limonite forms on the surface of rocks or as layers in the soil. Last, but certainly not least, goethite usually forms well-developed crystals that can look like needles made out of velvet, or thin prisms. In some cases, goethite can even form crystals that look like the stalactites that hang from the ceiling of caves, only much smaller. Limonite was always thought to be amorphous, meaning that it didn't form crystals at all. This all changed once detailed x-rays were taken of limonite and it was discovered that it did have a crystal structure that was the same as goethite, only really, really tiny.

Since you can actually see the crystals of goethite, it's also easy to measure some of its other properties. Goethite is fairly hard, but even though it contains iron, it does not feel heavier than most common minerals. Because it forms as the result of the weathering of other minerals, both goethite and limonite are found near Earth's surface. Not all goethite forms from the breakdown of other minerals, though. It can also form crystals directly from water that is rich in dissolved iron. As a result, chunks of goethite can be found in swamps and marshes where it is simply called bog iron. With all these different characteristics, it's no wonder that this iron oxide mineral has more than one name!

FACTS

CLASS: hydroxide minerals

MAIN COLORS: dark brown, yellowish brown, reddish brown

CLEAVAGE: one direction but also has an uneven fracture

LUSTER: silky or dull

STREAK: yellowish brown

HARDNESS: 4–5.5

OTHER DISTINCTIVE FEATURES: often found as nodules in bogs

THE PIGMENT **YELLOW OCHER** THAT IS USED TO MAKE CERTAIN **PAINTS** COMES FROM **GROUND-UP PIECES OF GOETHITE.**

URANINITE

Have you ever wondered where they get the fuel to run the reactors at nuclear power plants? Well, like so many other valuable resources that we depend on in our modern world, it starts with a rock that comes out of the ground.

Uraninite is an oxide mineral that is the most important source of the chemical element uranium, which is used to make the fuel for nuclear reactors. Nuclear reactors have several advantages over other ways of generating electricity. They don't create carbon dioxide, which gets into the air when fossil fuels such as coal, oil, or natural gas are burned. They are small, so they can fit inside of a submarine or even an uncrewed probe designed to fly into outer space. Of course, they do have one major drawback: Nuclear reactors produce radioactive waste, which can remain dangerous for thousands of years. And when things go wrong (like in a nuclear accident), they can create a major environmental disaster. For this reason, many countries have begun phasing out the use of nuclear energy.

Because it contains uranium, uraninite itself is radioactive and must be handled only by professionals and with extreme care. This is one of the minerals that you certainly don't want to add to your collection! Fortunately, it has a few properties that make it pretty easy to identify and avoid. First, it's almost always found in igneous rocks like granite pegmatites. While it can form crystals that are eight-sided octahedrons, it is most commonly found in a single mass or looking like a bunch of grapes forming a rock called pitchblende. Because uranium itself is a large chemical element, a chunk of uraninite feels heavier than most rocks. It's almost always a dull black color, although some samples may have a yellow coating on them. Of course, the most distinctive property is that it is radioactive. But you can't tell this just by looking at a rock. To tell if a rock is radioactive, geologists use a device called a Geiger counter, which detects and measures radiation.

FACTS

CLASS: oxide minerals

MAIN COLORS: pitch black, possibly with a yellow crust

CLEAVAGE: none; uneven fracture sometimes

LUSTER: submetallic or dull

STREAK: brownish black

HARDNESS: 5–6

OTHER DISTINCTIVE FEATURES: radioactive and very dense/heavy

THE **FIRST PLACE** THAT THE ELEMENT **HELIUM** WAS DISCOVERED **ON EARTH** WAS IN **URANINITE.** BEFORE THAT IT WAS ONLY KNOWN TO **EXIST** ON **THE SUN.**

Pitchblende is the most common form of the mineral uraninite, which is radioactive and can be extremely dangerous!

Bet You Didn't Know!

Scientists Marie and Pierre Curie performed groundbreaking work on radioactivity in the late 1800s. They used large quantities of pitchblende, from which they extracted uranium and eventually discovered the radioactive elements radium and polonium. Unfortunately, they didn't know about the dangers of radioactivity—Marie would even carry around tubes of radium in her pocket!

Chrysoberyl forms amazing crystals that sometimes even change colors.

One unusual form of chrysoberyl is called cat's eye because when it is polished and cut into the shape of a dome, it shines with a narrow band of light, just like the eye of a cat at night. The shimmering effect comes from tiny parallel spaces inside the crystal, which change the direction of the light rays as they bounce off.

CHRYSOBERYL

While gemstones like diamond (pp. 152–153), ruby (pp. 168–169), and sapphire (pp. 168–169) get lots of attention, there's one rare form of the mineral chrysoberyl called alexandrite that has them all beat. Not only does it sparkle beautifully when it is cut and polished, but it can actually change color from green to cherry red when you move it from daylight to artificial lighting indoors!

Even without this color-changing ability, the normal type of chrysoberyl is not a very common mineral. Chrysoberyl is made from a combination of beryllium, aluminum, and oxygen, and it usually forms six-sided, prism-shaped crystals. In some cases, the crystals will form heart-shaped twins with two crystals growing out from the same point. Much of the chrysoberyl that is sold in the world comes from Brazil, but it can also be found in Australia, Russia, Madagascar, Zimbabwe, and Myanmar (Burma).

Common forms of chrysoberyl are usually yellow, green, or brown and will often have fine lines called striations running parallel to the long side of the crystal faces. Because of the tight atomic structure, chrysoberyl is one of the hardest minerals, only being topped by diamond (pp. 152–153), moissanite, and corundum (pp. 166–167). It forms in both granite pegmatite and mica schist (p. 107) rocks, but because it is so hard, it does not break down easily. As a result, it is most often found mixed in with gravel that has been deposited by streams after it has weathered out of the rock that it formed in.

The name "chrysoberyl" is based on the Greek word *chryos,* meaning "golden beryl"; people used to think chrysoberyl was an unusual form of the mineral beryl (pp. 230–231), which is also quite hard and has long, six-sided crystals. Because of its unique properties, chrysoberyl has its own special place near the top of the list of outstanding gemstones.

FACTS

CLASS: oxide minerals

MAIN COLORS: yellow, green, brown

CLEAVAGE: distinct in one direction; uneven fracture sometimes conchoidal

LUSTER: glassy

STREAK: none

HARDNESS: 8.5

OTHER DISTINCTIVE FEATURES: may appear red under artificial light

Alexandrite appears green in daylight and red in indoor lighting.

DAYLIGHT

ARTIFICIAL LIGHT

ALEXANDRITE, THE **RARE** GEMSTONE VARIETY OF **CHRYSOBERYL,** WAS ORIGINALLY **FOUND IN RUSSIA** AND WAS NAMED FOR **ALEXANDER II, EMPEROR OF RUSSIA** FROM 1855 TO 1881.

SOME **REALLY** COOL **CRYSTALS**

Minerals that belong to the oxide and hydroxide classes have lots of practical uses, but sometimes they are just nice to look at. Here are five minerals that make some really cool-looking crystals—and one that can even cool off your drink on a hot summer day!

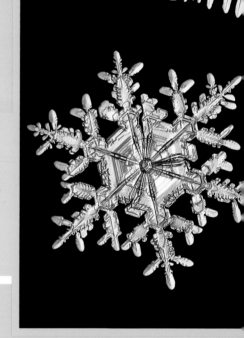

ICE

OK, we know what you're thinking: Ice can't be a mineral—it's just frozen water! But it actually has all the required properties of minerals, just like quartz (pp. 206–207) or feldspar (pp. 212–213). Ice is made of hydrogen and oxygen, so it is in the oxide class of minerals. When water freezes, it can form six-sided, or hexagonal, crystals, which is why snowflakes usually have six branches. The thing that sets ice apart from other minerals is that it has a really low melting point. Most other minerals don't even start to melt until they reach temperatures of hundreds or even thousands of degrees, but ice begins to melt as soon as the temperature reaches 32°F (0°C). Because a mineral has to be naturally occurring, the ice that you make in your freezer is not technically a mineral, but an iceberg in nature can be thought of as a really cold rock made from really cool minerals!

MANGANITE

Manganite is an important ore of the metal manganese, which is often added to steel to make it stronger. But in addition to this practical use, it can often form some spectacular bundles of prism-shaped crystals. Most often, the mineral is dark gray to jet black in color with a submetallic luster that can sometimes make the crystals shine. Manganite can be found in veins in a variety of rock types formed from hot, mineral-rich water flowing through them. Sometimes, this cool-looking mineral can even be found around hot springs.

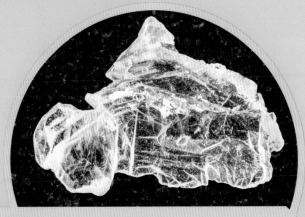

BRUCITE

Brucite is a soft hydroxide mineral that is used as a source of magnesium metal. It usually forms large transparent or translucent crystals that look like plates, but sometimes it almost looks like a bunch of waxy fibers that have been stuck together. In this form, the crystals can actually bend like a flexible mat! Brucite comes in a variety of colors including white, gray, bluish white, and pale green, and it's normally found in metamorphosed limestone (pp. 120–121), phyllite (p. 106), and schist (p. 107) rocks.

DIASPORE

Diaspore is a superhard hydroxide mineral made from the metal aluminum that is known for its thin crystals that look like little blades. Diaspore can come in a wide range of colors, including white, gray, yellow, pale green, and even pink, and it shines with a glassy or pearly luster. The mineral itself forms in metamorphic rocks such as schist (p. 107) and marble (pp. 100–101), and you can often find it with other oxide minerals that contain aluminum, such as spinel (pp. 170–171) and corundum (pp. 166–167). One rare form of diaspore is a transparent pink variety called zultanite, which is often cut and polished and used as a gemstone.

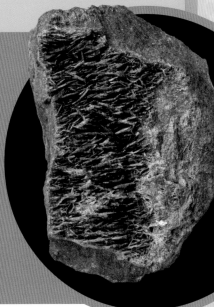

RUTILE

Rutile gets its name from the Latin word *rutilus*, which means "red," and it's a perfect description for the stunning, prism-shaped crystals that this mineral often produces. Rutile is an oxide mineral made from the metal titanium and is commonly found in granite (pp. 78–79), gneiss (p. 106), and mica schist (p. 107). Microscopic needlelike crystals of golden rutile can often be found inside clear quartz crystals (pp. 206–207), which can sometimes take the shape of a star!

SALT OF THE EARTH

Picture this: For miles in any direction, the only thing you can see is a gray-white plain that is flat as a pancake. In the far-off distance you can make out the pointed tops of high mountain peaks, but plants and animals are rarely found. When the wind blows, clouds of chalk-white dust fill the air, stinging your eyes and leaving a bitter taste in your mouth. Have you been abducted by aliens and dropped onto a dry, barren, and desolate world? Nope, this is no intergalactic landscape: It's Salar de Uyuni, the largest salt flat in the world.

A salt flat is a large plain that is covered by a thick layer of minerals that formed from the evaporation of salty water. As you might expect, many of the salt flats found in the world today are located near the ocean or close to sea level where saltwater is common, but Salar de Uyuni—located in the South American country of Bolivia—is the exact opposite. Sitting at almost 12,000 feet (3,658 m) above sea level, it's located right in between the eastern and western chains of the Andes Mountain range in an area called the Altiplano, or high plain. Like the Andes Mountains themselves, the Altiplano was raised by tectonic forces at work deep inside the Earth when the edge of the South American plate was pushed up in a collision with the Nasca plate under the Pacific Ocean. Many geologists believe this happened over the past 30 to 50 million years.

The salt flats are huge—covering over 4,000 square miles (10,000 sq km)—and are all that remain of a giant lake that covered

DURING THE **RAINY SEASON** AT **SALAR DE UYUNI,** THE SALT FLAT IS COVERED WITH A **THIN LAYER OF WATER** THAT **REFLECTS THE SKY** SO WELL THAT IT'S HARD TO KNOW **WHICH WAY IS UP!**

Bet You Didn't Know!

The salt that is found at Salar de Uyuni contains a large concentration of the element lithium, a key component in the long-life batteries used to power portable devices such as cell phones, laptop computers, and even some electric cars. Some scientists estimate that as much as half of the world's lithium reserves may be buried in Salar de Uyuni!

Bolivia's Salar de Uyuni is the world's largest salt flat, but it looks more like some distant planet!

A **PALACE** OF **SALT**

One of the most unusual hotels in the world is the Palacio de Sal, located right on the Uyuni salt flats. As you might have guessed, the "Palace of Salt" is not made from wood or concrete. Instead, it's made from more than 10,000 tons (9,072 t) of salt blocks cut right from the flats. Almost all of the furniture found in the hotel is also made of salt. We're talking tables, chairs, sofas, and even beds, all carved from giant blocks of salt.

There are a few drawbacks to having everything made from salt, though, especially when it rains. Because salt easily dissolves in water, some of the structures have to be replaced after the rainy season. Another problem is that salt can react with and destroy certain metals, a process known as corrosion, so sometimes electrical outlets and wiring stop working. And to protect the walls from extra wear and tear, there is a rule prohibiting guests from licking them!

the area between 18,500 and 8,500 years ago, when the climate was much wetter. As the water slowly evaporated, dissolved minerals in the water became more and more concentrated and began to form layers of crystals, eventually forming a thick crust of evaporite minerals including halite (pp. 186–187) and gypsum (pp. 200–201). People first started mining the salt back in the 1500s, and they are still at it today. A trip to Salar de Uyuni is as good as visiting another planet right here on Earth!

HALITE

You may not realize it, but at this very moment there are probably thousands of little bits of a halide mineral hanging out somewhere in your kitchen. In fact, unless you are really careful, you might even wind up eating some of it on your food! Don't worry, there is nothing to be afraid of, because the mineral we're talking about is called halite, and it's the scientific name for common table salt.

Halite is one of the most common minerals there is, and people have been using it to prepare foods for thousands of years. Not only does it give food a salty taste that many people enjoy, but long before there were refrigerators and freezers to keep food fresh, many types of meat were preserved by "salting" them down. Rubbing meat with salt helps to dry it out, which helps to slow the growth of bacteria and fungi that can make food spoil.

Halite forms when the chemical elements sodium and chlorine dissolve in water and join together to form cube-shaped crystals as the water slowly evaporates. While the most common forms of halite are either colorless or white, it can also come in pink, blue, gray, brown, and even orange. In the past, thick deposits of halite formed when large inland seas slowly dried up as the climate changed. Over time, these salt layers became buried, and the weight of sediments pressing down on them caused them to slowly flow up toward the surface. This resulted in the formation of large underground structures called salt domes, or salt diapirs. Today, you can find halite forming naturally in arid, or dry, areas where salty water is trapped in lakes and bays.

FACTS

CLASS: halide minerals

MAIN COLORS: clear, white, gray, orange, brown, pink, purple, blue

CLEAVAGE: perfect in three directions making square edges

LUSTER: glassy

STREAK: white

HARDNESS: 2.5

OTHER DISTINCTIVE FEATURES: salty taste, dissolves easily

BECAUSE IT IS SO **WIDE AND FLAT,** THE **BONNEVILLE SALT FLATS** IN THE U.S. STATE OF **UTAH** ARE OFTEN USED TO TEST OUT **HIGH-SPEED VEHICLES,** SOME OF WHICH HAVE SET **LAND SPEED RECORDS.**

Not all halite is white, but it almost always forms crystals shaped like cubes.

Bet You Didn't Know!

While some naturally forming halite is still used for food products, much of the salt that we eat comes from the evaporation of salty water called brine. This can either be done naturally using solar energy to evaporate the water or by heating the brine using some type of fuel such as oil or natural gas until the water evaporates.

Fluorite comes in a wide range of colors.

Bet You Didn't Know!

Some minerals, such as fluorite, tend to glow in a different color when they are placed under ultraviolet (UV) light, or black light. Fluorite, for example, will look blue under UV light. This special property is called fluorescence, and, as you may have guessed, the name comes from the mineral fluorite.

FLUORITE

They say that variety is the spice of life, and when it comes to mineral colors, fluorite is as spicy as it gets! Most minerals come in a few basic colors. There are some, such as quartz (pp. 206–207), that can come in as many as six different shades. But fluorite can come in more than a dozen colors, some of which have stripes.

Like halite (pp. 186–187), fluorite belongs to the halide class of minerals and is made from the elements calcium and fluorine. But it can also include some other chemical compounds in the mix, which is where all the different colors come from.

A very common mineral, fluorite can form in a wide range of rocks, including veins formed by mineral-rich fluids that have deposited ores of lead and silver. It also can be found in granite pegmatites, as well as in limestone (pp. 120–121) and in spaces found in other sedimentary rocks. It can even form around hot springs, where it is deposited by mineral-rich waters that flow out of the earth.

Fluorite forms crystals that look like cubes or eight-sided octahedrons. But in some cases, the crystals can be twins, with two crystals growing in opposite directions from the same starting point. The crystals are transparent or translucent and have a glassy luster. Most samples of fluorite are a single color, but some have distinct layers or stripes in them. One variety, called Blue John, has bands of purple, white, and sometimes yellow fluorite all in the same rock.

Fluorite is used in making many different products, including fuel, steel, and hydrofluoric acid, a solution so powerful it can dissolve glass. It is also used in making the enamel coating for pots and pans, and transparent varieties have even been turned into special types of lenses for cameras.

FACTS

CLASS: halide minerals

MAIN COLORS: violet, blue, green, yellow, pink, rose red, brown, bluish black, colorless, white, transparent

CLEAVAGE: perfect in four directions forming an octahedron, but with uneven fracture

LUSTER: glassy, dull

STREAK: white

HARDNESS: 4

OTHER DISTINCTIVE FEATURES: sometimes fluorescent under ultraviolet light

BLUE JOHN VARIETY OF FLUORITE

FLUORITE GETS ITS NAME FROM THE **LATIN WORD** *FLUERE,* WHICH MEANS **"TO FLOW,"** BECAUSE IT HAS A **LOW MELTING POINT** AND IS USED IN THE **SMELTING**— A PROCESS OF **HEATING** AND **MELTING**—OF CERTAIN METALS.

CALCITE

Calcite is one of the most important rock-forming minerals and is the most common form of the chemical compound calcium carbonate, which is also called lime. As you might have guessed, it's the main mineral found in many types of sedimentary rocks including limestone (pp. 120–121), chalk (pp. 122–123), and travertine (pp. 118–119), which is formed in hot springs or in stalactites in caves. It is the main mineral in marble (pp. 100–101), too, which is usually formed when limestones are metamorphosed. While it is not common, calcite can also be found in both mafic and felsic igneous rocks including pegmatites and basalt (p. 90), especially when they intrude through other rocks. Rocks containing calcite have many uses and are mined all around the world. It's the main ingredient in cement, is used in making steel and other metals, and farmers use powdered calcite to add nutrients to the soil.

Calcite can form a wide range of crystal shapes depending on the environment where it is found, or it can be massive and not form any large crystals, as is the case with most limestone and marble. When it does form crystals, they can be quite large and beautifully shaped. Two of calcite's common crystal forms are rhombohedrons, which look a little like bent cubes, and 12-sided scalenohedra, which look like pyramids made from uneven-sided triangles. Calcite crystals can also form flat plates and be twins, which makes them look like the wings of a butterfly!

Calcite has a number of physical properties that make it easy to identify in rocks, even when it doesn't form well-developed crystals. It is soft enough to be scratched with a penny, and it usually has a glassy luster. Calcite is quite brittle, and when you hit a chunk it will almost always break into little rhombohedrons with edges looking like bent cubes. Finally, calcite produces a chemical reaction when acid is placed on it, releasing carbon dioxide gas. This means that if you put a drop of vinegar on calcite, it will begin to fizz like seltzer!

FACTS

CLASS: carbonate minerals

MAIN COLORS: usually white or colorless but can be pale gray, yellow, red, or green; brown or black if not pure

CLEAVAGE: distinct—breaks into rhombohedrons

LUSTER: glassy to earthy, transparent to translucent

STREAK: white

HARDNESS: 3

OTHER DISTINCTIVE FEATURES: fizzes when an acid such as vinegar is placed on it

MANY FORMS OF **CALCITE** ARE CALLED **SPAR,** WHICH IS AN **OLD TERM** THAT **MINERS USED** IN THE PAST FOR **LIGHT-COLORED MINERALS** THAT HAD **DISTINCT CLEAVAGE** AND WERE SHINY **LIKE GLASS.**

The mineral calcite doesn't always form crystals, but when it does, they are often spectacular and in the shape of 12-sided scalenohedra.

ICELANDIC SPAR

Bet You Didn't Know!

Crystal clear pieces of Icelandic spar calcite have a property called double refraction. This means that when you look at something through the crystal, it creates two images of the object you are looking at. Even though it is named for Iceland, this unusual type of calcite can form pretty much anywhere in the world.

SOME SERIOUSLY CLASSY CARBONATES!

Carbonate minerals such as calcite (pp. 190–191) and dolomite (pp. 194–195) are important mineral resources, but there are a few others that people go out of their way to collect because of their unusual colors, patterns, or crystal shapes. These minerals are frequently cut and polished to be used for jewelry and decorative items such as bookends. Here are five carbonate minerals with some real "class" that collectors adore.

MALACHITE

If green is your color, then malachite is the mineral for you. Just like the Statue of Liberty, malachite gets its green coloring from the element copper, and people have been using it as a copper ore for close to 5,000 years. Malachite rarely forms crystals. The thing that collectors are looking for is the amazing swirling patterns that form in the mineral. Malachite is a secondary mineral, which means that it forms from other minerals that have been changed by chemical reactions after they have been deposited. Malachite is most commonly found as a solid mass or as a crust on the outside edges of other rocks that have been exposed to air and flowing water and that have copper-forming minerals in them. When the malachite is cut and polished, it is used for all sorts of decorative pieces, including bowls, dishes, and vases. It can even be used as a gemstone.

AZURITE

Like malachite, azurite is a secondary mineral that forms when other copper-rich minerals go through a chemical change, and the two are often found together. A major difference between the two is that azurite is a deep shade of blue instead of green. This mineral will often form as a mass, but it can also form some really nice prismatic and platelike crystals. When it does form as massive chunks, it is sometimes called chessylite, named for Chessy, France, where it is very common. Along with polishing azurite to create decorative stones, artists during the 15th, 16th, and 17th centuries also ground it up to make the deep azure blue pigment used in their paintings.

CERUSSITE

Let's face it: Compared with gold, silver, or even copper, lead is not usually thought of as being an attractive metal. But cerussite takes this superdense element to a whole new level. Lead-based minerals such as galena (pp. 158–159) tend to be dull and blocky, but cerussite is the exact opposite, shining with a sparkly luster that reflects light off its different crystal faces. Cerussite is a secondary mineral that forms when carbonated water (which is basically natural seltzer) reacts with galena to produce delicate crystals that can be shaped like prisms or plates. A favorite of collectors is called jack-straw cerussite, pictured here, which looks like a bunch of needles.

RHODOCHROSITE

Rhodochrosite is a carbonate mineral that forms from the chemical element manganese, which helps to produce the bright rose pink color that is loved by many mineral collectors. Like calcite (pp. 190–191), rhodochrosite can form rhombohedron crystals, but these are rare, which makes them more valuable than other types of rhodochrosite. Along with galena (pp. 158–159) and sphalerite (p. 165), rhodochrosite is often found in veins of metallic minerals deposited by mineral-rich fluids that flowed through cracks in other rocks. Rhodochrosite gets its unusual name from the Greek term *rhodokhros,* which means "of rosy color," and while it can come in brown and gray, the really valuable pieces live up to their rosy name.

SIDERITE

Most common types of iron ore, like hematite (p. 174) and magnetite (pp. 172–173), are massive minerals that are not very exciting to look at. Siderite, on the other hand, is known as much for the wide range of crystals it can form as it is for its "heavy metal" content. Siderite is found in all three major rock types, and its crystals can be rhombohedrons, platelike, prisms, or even 12-sided pyramids. Most often they are either light or dark brown with a glassy or pearly luster, and some can even be transparent or translucent. Now that's a different type of iron ore for sure!

Dolomite crystals look similar to those of calcite, except the faces are usually curved, giving them a saddle shape.

Bet You Didn't Know!

Not too long ago, if geologists were talking about dolomite you didn't know whether they were talking about the mineral or the rock that is made up mainly of the mineral dolomite. They both had the exact same name. To solve the confusion, the name "dolostone" is now used to describe the rock, and the name "dolomite" is used just for the mineral.

DOLOMITE

Calcite (pp. 190–191) and dolomite have a very close relationship. Not only are they both important rock-forming carbonate minerals, but they share many of the same properties, which often makes it difficult to tell them apart. The reason for their similarities is that they have almost the same chemical composition. While calcite is pure calcium carbonate, dolomite is calcium carbonate with some of the element magnesium thrown into the mix. That extra magnesium is just enough to change the internal structure of the mineral, and it usually comes from magnesium-rich water flowing through limestones made of calcite. When this happens, the calcite changes to dolomite, and the limestone changes to a new rock called dolostone. This process is sometimes called dolomitization.

Like calcite, dolomite is most often found as a massive grayish white mineral in both layered sedimentary rocks and marbles (pp. 100–101) that formed when dolostones underwent metamorphism. Sometimes large crystals of dolomite can be found growing in the open spaces in these rocks, and they can also be found in veins that contain lead and zinc minerals. Like calcite, dolomite crystals are mostly shaped like rhombohedrons that look like bent cubes, but the faces are more curved, so they can often be saddle-shaped. In some cases, dolomite will appear as small bunches of tabular, platelike crystals that form a crust on the surface of another rock.

Since they are so similar, the easiest way to tell dolomite and calcite apart is with the "acid test." If you put a drop of vinegar on calcite, it will begin to fizz right away. To get dolomite to fizz, you either need to scratch the surface of the crystal first or crush it into a powder, and even then the reaction will be much slower than with calcite. Of course, this test isn't always 100 percent accurate. Why? Since many rocks that have dolomite in them still have some calcite too, it could be the calcite that's doing the fizzing.

FACTS

CLASS: carbonate minerals

MAIN COLORS: colorless, white, gray, pink; less often green, brown, black

CLEAVAGE: perfect in three directions forming rhombohedrons

LUSTER: glassy

STREAK: white

HARDNESS: 3.5–4

OTHER DISTINCTIVE FEATURES: will fizz slightly with vinegar after being scratched

THE MINERAL **DOLOMITE** WAS **NAMED** IN HONOR OF THE **FRENCH GEOLOGIST DIEUDONNÉ DOLOMIEU,** WHO WAS THE FIRST PERSON TO DO A DETAILED **GEOLOGIC STUDY** OF THE MINERAL IN THE LATE **1700S.**

BORAX

Death Valley in the U.S. state of California is one of the most inhospitable places on the planet. Located near the border of Nevada, it is part of the Great Basin, which is a low-lying area in the western United States between the Sierra Nevada and Wasatch mountain ranges. At 282 feet (86 m) below sea level, it has the lowest ground elevation in North America. During summer, temperatures have climbed to more than 130°F (54.4°C), making it one of the hottest, driest places on Earth. One thing that Death Valley has going for it is that it is a source of many minerals including gold (pp. 142–143), silver (pp. 146–147), and especially borax.

Borax, like halite (pp. 186–187), mostly forms as a sedimentary mineral when mineral-rich water that is trapped in an enclosed basin like Death Valley slowly evaporates. Today there is very little water to be found in Death Valley, but near the end of the last ice age much of the area was covered by a large lake, similar to the Great Salt Lake in the U.S. state of Utah today. Earlier volcanic activity and periodic floods washed mud and minerals into the lake, especially the chemical element boron, which was in the volcanic ash.

Borax often forms thick, massive beds, but sometimes it will create short prism-shaped crystals that grow on the surrounding rocks. When the crystals first form, they are usually colorless and transparent, looking a little like glassy needles. As soon as the borax crystals are exposed to the dry air, the mineral's surface begins changing into a substance called tincalconite, which makes it look and feel like chalk (pp. 122–123).

Borax is quite soft and can be scratched with the edge of a penny. It also dissolves easily in water. The main use of borax is as a source of the element boron, which is used in the making of different metals. Borax is also used as a disinfectant and a laundry detergent, and it can be found in paints and even ceramic pottery. Considering that it comes from a place called Death Valley, borax is actually some pretty lively stuff!

FACTS

CLASS: borate minerals

MAIN COLORS: colorless or white

CLEAVAGE: perfect but will also show conchoidal fracture

LUSTER: glassy or earthy

STREAK: white

HARDNESS: 2–2.5

OTHER DISTINCTIVE FEATURES: dissolves in water

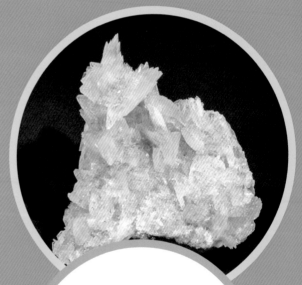

MINING **NATURAL BORAX** FROM PLACES LIKE **DEATH VALLEY** ISN'T THE ONLY WAY TO GET THIS MINERAL. IT CAN ALSO BE **MADE BY HEATING** A MINERAL CALLED **COLEMANITE** TO A **HIGH TEMPERATURE.**

Borax usually forms in thick, massive layers, but it can sometimes form crystals that look like stubby little prisms. The crystals start out clear but get a white powdery covering after they have been exposed to the air.

Bet You Didn't Know!
After borax was discovered in Death Valley in the late 1800s, it was hauled out in wagons pulled by up to 20 mules hitched together. The people who drove the wagons were called teamsters, which is the name of a union that includes truck drivers today. One popular brand of cleaner that you can still buy today is called 20 Mule Team Borax.

Even though it rarely forms crystals, turquoise has long been used as a gemstone because of its spectacular color.

Bet You Didn't Know!

Turquoise has a special place in Native American culture in the southwestern United States. Native Americans have been collecting and using it for more than 1,200 years. In some Native cultures, turquoise is used in art and jewelry. Its special color represents water and the sky as well as health and protection.

TURQUOISE

Many gemstones, such as rubies (pp. 168–169), emeralds (pp. 230–231), and diamonds (pp. 152–153), are prized for their beautiful crystal faces and the way that they both bounce and bend the light that hits them. Turquoise has none of these properties, yet because of its special color it was still one of the earliest stones to be intentionally collected, cut, and polished to be used for jewelry.

Archaeologists working in the Middle East have found beads and carved figures made from turquoise dating back more than 7,000 years, and it is still one of the most popular gemstones in use today.

The thing that sets turquoise apart from other ornamental stones is its blue-green color, which comes from the copper found in its mineral structure. While turquoise can form crystals, they are quite rare. Most often, it can be found filling in veins or as nodules in volcanic rocks in dry environments. Turquoise is called a secondary mineral because it forms from other copper-containing minerals that have been changed by substances found in the surrounding environment. In addition to its distinct color, turquoise has a waxy or even dull luster. It only really shines brightly when it is highly polished.

Even though it has a distinctive color, turquoise can be confused for other blue-green minerals, especially one called variscite, which also makes nodules. One way to tell them apart is that variscite tends to be greener in color and has a hardness rating on the Mohs scale of 3.5 to 4.5, making it softer than turquoise. Turquoise can be porous or have lots of little holes in the surface. Because of this, some turquoise that is sold may be "doctored," meaning the holes have been filled with plastic resin to make the gemstone shinier. In some cases, poor-quality turquoise samples are even dyed to improve their color. Of course, the best way to get real turquoise is to collect it on your own as people have been doing for thousands of years.

FACTS

CLASS: phosphate minerals

MAIN COLORS: blue, bluish green, green

CLEAVAGE: uneven cleavage but with conchoidal fracture

LUSTER: waxy

STREAK: white or pale green

HARDNESS: 5–6

OTHER DISTINCTIVE FEATURES: can be polished to a high shine

THE COLORING OF THE MINERAL **TURQUOISE** IS SO **UNIQUE** THAT THE NAME IS ACTUALLY USED TO **DESCRIBE** A **BLUISH GREEN COLOR** USED IN **PAINTS,** COLORED PENCILS, AND CRAYONS.

GYPSUM

Gypsum is a very common mineral, but it can take on some very uncommon shapes depending on the conditions under which it forms. Like many other sulfate minerals, it is soft, light in color, and the crystals are frequently transparent or translucent.

Most types of gypsum form as evaporite deposits when lakes that are rich in mineral salts dry out. When this happens, the gypsum forms thick, massive beds mixed in with layers of other evaporite minerals including halite (pp. 186–187). It may also be mixed with limestone (pp. 120–121) and shale (p. 127). These types of deposits are common in many areas of the world that were once arid but are no longer, and they help show how the climate has changed over time.

Under certain conditions gypsum can take on a variety of crystal forms including plates, blades that resemble swords, and crystal twins that are known as fishtails. In some cases, gypsum crystals will grow around pieces of quartz sand, forming a shape called a desert rose that looks like the bud of a flower. And then there's satin spar, which looks a little like bundles of fibers and forms in veins along with sulfide minerals such as galena (pp. 158–159).

While gypsum crystals can be attractive and fun to collect, this mineral is also a major resource, especially for construction. Gypsum is ground up and heated to remove the water it contains to create gypsum powder, which is used to make plaster of paris. When mixed with water, plaster of paris can be spread on walls and molded into different shapes. When it dries, it turns rock-hard. Gypsum is also used to make large boards called Sheetrock, which are used to construct interior walls and ceilings in most modern homes and buildings. Finally, blocks of massive white fine-grained gypsum called alabaster are cut and used to create decorative stones on the outside of buildings. Yes, for a fairly common mineral, gypsum has some pretty uncommon credentials!

FACTS

CLASS: sulfate minerals

MAIN COLORS: colorless, white, gray, yellow, brown; often transparent or translucent

CLEAVAGE: perfect in three directions, but with fibrous, almost splinter-like fracture

LUSTER: glassy, pearly and silky

STREAK: white

HARDNESS: 2

OTHER DISTINCTIVE FEATURES: often splits into thin strips

WHEN IT **FIRST FORMS, SELENITE GYPSUM** IS **TOTALLY TRANSPARENT** AND ALMOST LOOKS LIKE **GLASS,** BUT BECAUSE IT IS **SO SOFT,** IT EASILY **SCRATCHES.** AFTER BEING **EXPOSED** TO THE **ATMOSPHERE** FOR A WHILE IT BEGINS TO **TURN WHITE.**

The mineral gypsum can take on many different forms, including pure white alabaster, see-through selenite, and crystals that look like flowers.

Bet You Didn't Know!

White Sands National Park in the U.S. state of New Mexico is a unique environment. From a distance, it looks like the area is covered by sand dunes like those you'd find in many other arid areas. But if you take a close look at the sand, you'll see a difference. Instead of being made of quartz (pp. 206–207), all of the dunes are made of pure white particles of a crystallized form of gypsum called selenite.

CAVE OF THE CRYSTALS

The first thing that many people ask when they see scenes from inside Mexico's Cueva de los Cristales, or Cave of the Crystals, is whether the pictures have been photoshopped. After all, it's not every day that you see images of people crawling over gypsum (pp. 200–201) crystals so large that they make the people look like tiny insects. To say that the Cave of the Crystals is a unique environment would be a huge understatement. So far, scientists have not found any other place quite like it on Earth.

In the year 2000, two local miners discovered the cave, which lies about 1,000 feet (300 m) under the surface and is located beneath Sierra de Naica Mountain in the Mexican state of Chihuahua. The cave itself is a horseshoe-shaped chamber that is about 100 feet (30 m) long by about 30 feet (10 m) wide, roughly the size of a narrow basketball court.

The crystals, however, are off the chart in size. Made of a type of gypsum called selenite, the largest are up to 36 feet (11 m) long and weigh a whopping 55 tons (50 t). They have grown to this enormous size due to the special conditions in the cave. About 26 million years ago, the volcanic activity that created the mountain also deposited a large amount of a mineral called anhydrite in the rocks below the mountain. Anhydrite is made of a compound called calcium sulfate and has the same basic chemical composition as gypsum, but there is no water in its crystal structure. Unlike gypsum, it needs to be kept at high temperatures or it begins to dissolve, which is what happened inside the cave.

As the anhydrite dissolved, the water inside the cave became enriched in calcium sulfate, and new gypsum crystals, which can grow at lower temperatures, began to form. It turns out that there is a pool of magma underneath the cave that kept the water at just the right temperature to allow the gypsum crystals to keep growing and growing.

Bet You Didn't Know!

Even with the water removed, exploring the Cave of the Crystals was dangerous business. The hot pool of magma below the cave kept the temperature at a toasty 113°F (47°C), and the air was wet and heavy to breathe. As a result, visitors who planned on staying inside for more than a few minutes needed to wear special "cold suits" and respirators so they could breathe without damaging their lungs.

People look like ants crawling on the giant gypsum crystals inside Mexico's Cave of the Crystals.

TO **PUMP** OR **NOT** TO PUMP

Scientists estimate that the selenite gypsum crystals found in Cueva de los Cristales have been actively growing for at least half a million years. Researchers wanted to study the huge crystals, but in order to access the cave, all of the groundwater had to be pumped out. Once the water was removed from the cave, the crystals stopped growing.

After a few years, of being water free, scientists started to notice that some of the crystals in the cave were showing signs of breaking down and began to wonder what the long-term impact of pumping would be. In 2017, the decision was made to turn off the pumps and allow the cave to fill with water again. Not only should this preserve the crystals, but they might even start growing again. While it means that people won't be able to visit the cave, we'll always have the pictures.

BEFORE THE **CAVE OF THE CRYSTALS** WAS DISCOVERED, MINERS WORKING IN THE SAME AREA DISCOVERED **THE CAVE OF THE SWORDS** (BELOW), WHICH IS FILLED WITH **DAGGER-SHAPED SELENITE CRYSTALS.** THOUGH THESE CRYSTALS WERE LARGE, AT ABOUT **EIGHT FEET** (2.5 M) LONG, THEY WERE **MUCH SMALLER** THAN THOSE IN THE **CAVE OF THE CRYSTALS!**

SUPER SILICATES!

QUARTZ

If you were to analyze the chemical composition of Earth's crust, you would find that a little more than three-quarters of it (by weight) is made up of only two elements: oxygen and silicon. Based on this information, you would probably guess that many of the minerals that we find in the crust would also have these two elements in them, and you'd be correct! They're called silicates, and they make up the largest class of minerals on Earth. Along with carbonate minerals, silicate minerals are some of the most important when it comes to forming rocks. Because there are literally hundreds of different individual silicate minerals, geologists classify, or organize, them based on their similar crystal structures and chemical composition. Here's a rundown of some of the major silicate minerals and the types of rocks they are commonly found in.

QUARTZ is one of the most common minerals on the planet, showing up in all rock types, often in great abundance. Because it is very resistant to weathering, quartz is usually found in most types of sandstone (p. 127).

THE FELDSPARS are usually divided into two groups depending on their chemical composition. The plagioclase feldspars

MOST OF THE **MINERALS** FOUND IN THE **PLANTS WE EAT** COME FROM **SILICATE MINERALS** IN THE **SOIL** THAT GIVE PLANTS THEIR **NUTRIENTS.**

OLIVINE

Bet You Didn't Know!

Silicate minerals play many important roles in our day-to-day lives. They are found in most of the building materials that we use including concrete, stone, brick, and even glass. Without them there would be very little sand to play in at the beach or clay to make pottery and those cool sculptures in art class at school!

contain the elements calcium and sodium and are found in high concentrations in many igneous and metamorphic rocks. The same can be said for the pinkish potassic feldspars, which contain the element potassium.

THE MICA FAMILY of minerals are common in both mafic and felsic igneous rocks and are major components of many metamorphic rocks, especially schists (p. 107).

THE PYROXENE FAMILY usually includes dark-colored minerals found in both mafic igneous rocks such as basalt (p. 90) and several types of metamorphic rock.

THE AMPHIBOLE FAMILY, like pyroxenes, is a major player in the formation of mafic igneous rocks as well as metamorphic rocks, especially amphibolite (pp. 104–105).

THE OLIVINE GROUP are usually dark-colored minerals including forsterite and fayalite. Olivine minerals are rich in the elements iron or magnesium, and sometimes they are rich in both. They do the "heavy lifting" when it comes to forming mafic igneous rocks.

THE GARNET FAMILY includes minerals such as almandine and pyrope, which are found in some igneous rocks, but they are major players when it comes to forming metamorphic rocks.

HORNBLENDE

Even though quartz, olivine, and hornblende look very different from one another, they all belong to a group of minerals called silicates because they are all made from the same basic chemical elements.

THE **POWER** OF THE **PYRAMID**

The basic building block of all silicate minerals is the silicon tetrahedron. You can think of this structure as a pyramid made of four large oxygen atoms forming the four corners and a small silicon atom stuck in the middle. Just like the building blocks that kids play with, these silicon tetrahedra can join together in many different ways, with each structure forming a different silicate mineral.

For example, in the pyroxenes and amphiboles, the tetrahedra form long chains, while in the mica minerals they make large, flat sheets that stack on top of each other. Both quartz and feldspar have a complex arrangement of tetrahedra in three dimensions. It's amazing just how many different minerals can come from one simple building block. Just goes to show that there is a lot of power in that little pyramid!

SILICON TETRAHEDRON

205

Quartz, which has crystals that can be microscopic or incredibly large, is one of the most common minerals on Earth.

Bet You Didn't Know!

Macrocrystalline quartz has many different names, depending on its color and clarity. Some of these include the gemstone amethyst, which is purple; citrine, which is brownish yellow or yellow-orange; rose quartz, which is pink; and smoky quartz, which is dark gray to black. Then there is pure white milky quartz. And finally, there's the crystal-clear variety that's simply called rock crystal.

QUARTZ

If you asked 10 geologists to list the five most important rock-forming minerals, you'd almost certainly find quartz on everybody's list. Because of its unique set of properties and the many different ways it can form, quartz is truly a most valuable player when it comes to making rocks.

Quartz is actually the second most abundant mineral found in Earth's crust. The honor of being number one goes to the feldspars (pp. 212–213). Even though it comes in dozens of different forms, quartz has two main varieties. Macrocrystalline quartz forms crystals that are large enough to be seen with the naked eye. These crystals are almost always hexagonal, or six-sided, prisms—although sometimes they're cut off so that they look like pyramids instead.

Microcrystalline quartz forms a tight network of crystals that are so small you need a microscope to see them. This type of quartz usually has specialized names depending on the colors or patterns found in the rock. The general name is chalcedony, but varieties with circular bands are called agate, and the term "jasper" is used for a reddish variety that is stained by other mineral impurities.

One of the reasons why quartz is so common is because it can form in all three major rock types. It is a major component of felsic igneous rocks, especially granite (pp. 78–79) and rhyolite (p. 90). It can also be found in many metamorphic rocks, including quartzite (pp. 102–103) and gneiss (p. 106). Because quartz is very resistant to weathering, it is also a major component of coarse-grained clastic sedimentary rocks like sandstone (p. 127) and conglomerates. If that weren't enough, quartz can also form as a secondary mineral when water containing high amounts of dissolved quartz known as silica flows through openings in rocks and fills them with either macrocrystalline or microcrystalline varieties. Yes, when it comes to rock-forming minerals, quartz in all its varieties is a true standout!

FACTS

CLASS: silicate minerals

MAIN COLORS: clear, white, pink, purple, brown, yellow, gray, black

CLEAVAGE: rare and indistinct; shows good conchoidal fracture

LUSTER: glassy

STREAK: white

HARDNESS: 7

OTHER DISTINCTIVE FEATURES: hardness and hexagonal crystals

PEOPLE IN **ANCIENT GREECE** CALLED LARGE TRANSPARENT **QUARTZ CRYSTALS** *KRYSTALLOS,* WHICH IS WHERE THE **MODERN WORD "CRYSTAL"** COMES FROM.

NATURE'S **SURPRISE** PACKAGES

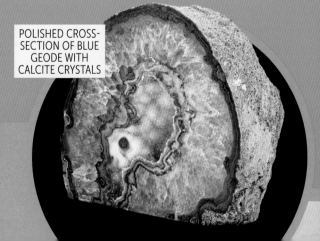

I magine that you are sitting around your home reading this book when the doorbell rings. It's your friend with a birthday present for you! It's in a big box with no markings on it that has been taped shut. After much struggle you finally manage to rip the box open, and to your amazement there is a giant rock filled with purple amethyst crystals inside. This is pretty much what happens every time a person cracks open a geode, nature's ultimate surprise package!

A geode is a lump of rock that has a roundish shape and is hollow on the inside. On the outside it's a dull shade of brown or gray with a rough, bumpy texture. Once you crack one open, that's when the fun begins. Many geodes contain spectacular crystals with the most common being some variety of quartz. Of course, sometimes instead of large crystals, there is a layer of bubbly looking micro-crystalline quartz, called chalcedony. In chalcedony, the crystals are so tiny you need a microscope to see them! There is really no way of knowing what you are going to get until you open the geodes up.

Geodes usually form in either sedimentary rocks (mostly limestones) (pp. 120–121) or ancient volcanic lava flows. The key ingredient is that the rock has to have big holes in it. In sedimentary rocks, these holes can come from tree roots that have rotted away or burrows that animals have dug. In volcanic rocks, the holes are the result of gas bubbles being trapped in the lava as it cooled. Think pumice (pp. 88–89), only with much bigger holes.

AMETHYST GEODES VS. AGATE NODULES

Because geodes can have lots of different mineral matter filling them, there is often a great deal of confusion about what they should be called. The official name for a round, hollow rock that has large crystals or a layer of the smaller chalcedony crystals inside is geode. Depending on what the crystals are made of, the name of that material will be added to the name of the geode. For example, if the crystals are made of amethyst, then it is an amethyst geode. If the crystals have calcite, then it is a calcite geode.

Sometimes when minerals fill in the hole in the host rock, instead of large crystals, microcrystalline quartz completely fills the space. When this happens, the rock is called a nodule. If the quartz inside is layered, it is called an agate nodule. And if it is a solid red color, it's usually called a jasper nodule. Basically, a geode and a nodule are the same thing except that nodules are completely filled in, while geodes still have a hollow space in the middle.

POLISHED CROSS-SECTION OF BLUE GEODE WITH CALCITE CRYSTALS

The type of mineral that forms inside a geode depends on the chemistry of the water that is flowing through the rock. Geodes can be filled with crystals of calcite (pp. 190–191), dolomite (pp. 194–195), pyrite (pp. 160–161), and even more exotic minerals such as sphalerite (p. 165).

Geodes can have spectacular crystals. When they become completely filled with micro-crystalline quartz, they are called agate nodules (see below).

AMETHYST GEODE

BECAUSE OF THEIR **ROUNDED SHAPE, GEODES** GET THEIR **NAME** FROM THE **GREEK TERM GEOIDES,** WHICH MEANS **"EARTHLIKE."**

CUT SECTIONS OF AGATE NODULES

As water flows through the rock, it dissolves some of the minerals and carries them away. When this mineral-rich water reaches one of the holes, new minerals begin to grow along the inside edge of the space. Over time, this mineral crust gets thicker and thicker, filling the hole from the outside in. Sometimes the minerals form crystals and sometimes they don't. After thousands or sometimes millions of years of weathering and erosion, the surrounding rock, which is usually softer than the new minerals filling the space, wears away and you're left with a round hunk of rock that is called a geode. And it's just waiting to be cracked open to reveal the surprise inside!

OPAL

Here's a riddle for you: Which rock is made up of the same chemical elements as quartz, has the same colors of quartz, forms in many of the same environments as quartz, and even looks a little like microcrystalline quartz, but isn't quartz? The answer is opal, which is what you get when you take the same basic ingredients as quartz and add a little bit of water into the mix. Many references list opal as a silicate mineral, but the truth is it is a mineraloid—a mineral-like material that does not have an internal crystal structure. Technically, opal can't be classified as a mineral.

What opal lacks in crystals, it more than makes up for in luster. Basically, opal is made from tiny spheres of silica separated by water molecules. This gives opal a unique property called play of colors. When light hits the little silica spheres, it bounces around creating a whole range of colors including red, orange, green, and blue. The effect, called opalescence, is dazzling and makes opal a very popular gemstone.

Opal comes in several varieties and only a few—especially "precious opal"—have a dramatic opalescence. Other types of opal include "common opal," which can come in a variety of colors but does not have any opalescence; black opals, which can also be dark blue or dark green; and fire opals, which, as the name suggests, are red or orange and seem to flash when light hits it.

Like quartz, opal can form in all three major rock types where it fills in openings after being deposited by water that is rich in silica. It is especially common in basalt (p. 90) and rhyolite (p. 90) lava flows and can form as nodules, veins, or even structures that look like stalactites. Since it contains water in its structure, opal can literally dry out and begin to develop tiny cracks after it solidifies. Because of this, uncut opal gemstones are often stored in water.

FACTS

CLASS: listed as a silicate mineral even though it is a mineraloid

MAIN COLORS: colorless, white, yellow, orange, red, purple, blue, green, gray, brown, black

CLEAVAGE: none; conchoidal fracture common

LUSTER: glassy, waxy, greasy, dull

STREAK: white

HARDNESS: 5.5–6.5

OTHER DISTINCTIVE FEATURES: appears to change color when the light hits it from different directions

OPAL MAY HAVE GOTTEN ITS **NAME** FROM THE **ANCIENT SANSKRIT WORD UPALA,** WHICH MEANS **"PRECIOUS STONE."** OPAL IS ALSO ONE OF THE **BIRTHSTONES** FOR THE MONTH OF **OCTOBER.**

The mineraloid opal shows a unique "play of colors" when white light hits its surface.

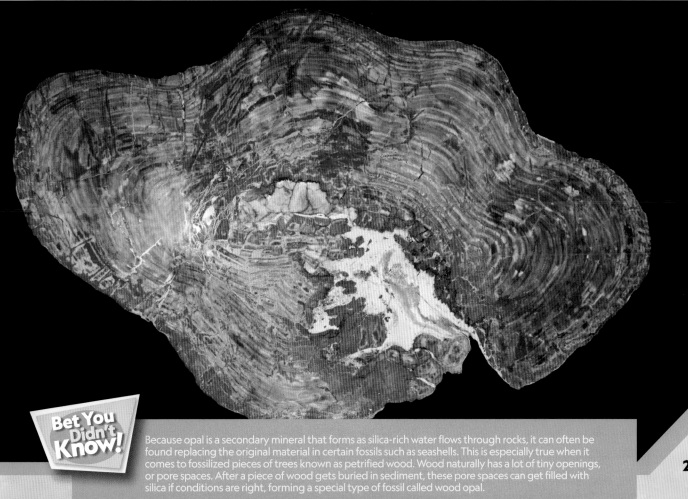

Because opal is a secondary mineral that forms as silica-rich water flows through rocks, it can often be found replacing the original material in certain fossils such as seashells. This is especially true when it comes to fossilized pieces of trees known as petrified wood. Wood naturally has a lot of tiny openings, or pore spaces. After a piece of wood gets buried in sediment, these pore spaces can get filled with silica if conditions are right, forming a special type of fossil called wood opal.

Bet You Didn't Know!

PHENOMENAL
FELDSPARS!

As a group, the feldspars make up some of the most important minerals in Earth's crust. Feldspars are easy to recognize because they are all relatively hard (6+ on the Mohs scale). And they have really good cleavage in at least two directions, so the minerals often break leaving sharp edges on them. They are usually divided into two groups based on their chemical composition. Plagioclase feldspars contain different amounts of calcium and sodium, while potassic feldspars are all rich in potassium. Here's a look at some of the most common rock-forming feldspar minerals.

MICROCLINE

Along with sanidine and orthoclase, microcline is one of the three major potassic feldspars. It can be white, salmon pink, or green and can form some incredibly large prism-shaped crystals in a type of felsic igneous rock known as a pegmatite. These crystals often weigh several tons! When microcline forms with quartz crystals (pp. 206–207) in "graphic granite," it can make an interesting pattern that almost looks like writing. One blue-green variety of microcline called Amazonite (right) is polished and used as a gemstone. It is also ground up and used in making ceramics.

ALBITE

Albite is a plagioclase feldspar that is rich in the chemical element sodium and is most commonly found in felsic igneous rocks, including granite pegmatites and many regional metamorphic rocks such as schist (p. 107). Albite usually forms tabular or platelike crystals and can be colorless, pink, yellow, or green but is most often white and gets its name from the Latin word *albus,* which means "white."

SANIDINE

Sanidine is a potassic feldspar that is very similar to both orthoclase and microcline, except that it usually forms at high temperatures. It is a common component in felsic volcanic rocks such as rhyolite (p. 90) and can be found in some metamorphic rocks that have been under low-pressure but high-temperature conditions. Sanidine usually forms flat, tabular crystals, which can sometimes form twins with two crystals growing together in opposite directions. Sometimes it can be found as tiny needle-like crystals in obsidian (pp. 86–87), which creates a snowflake effect.

ORTHOCLASE

Orthoclase is a potassic feldspar that forms prism-shaped crystals that come in a variety of colors including white, pink, brown, gray, and green, and it can also be colorless. It is common in intrusive igneous rocks such as granite (pp. 78–79) that cooled slowly from a felsic magma and can also be found in metamorphic gneiss (p. 106). One variety, called moonstone, has the same type of play of colors as opal (pp. 210–211) and is used as a gemstone.

ANORTHITE

Anorthite is similar to albite, but instead of being made from the element sodium it is rich in calcium. It can be found in mafic igneous volcanic rocks such as basalt (p. 90), is common in marbles (pp. 100–101) that formed when limestones underwent contact metamorphism, and has even been found in some meteorites! Anorthite usually forms white, pink, red, or gray prism-shaped crystals that appear to be bent at a strange angle, which is how it got its name. *Anorthos* is a Greek term that basically means "not straight."

COOKING UP SOME FELDSPATHOIDS

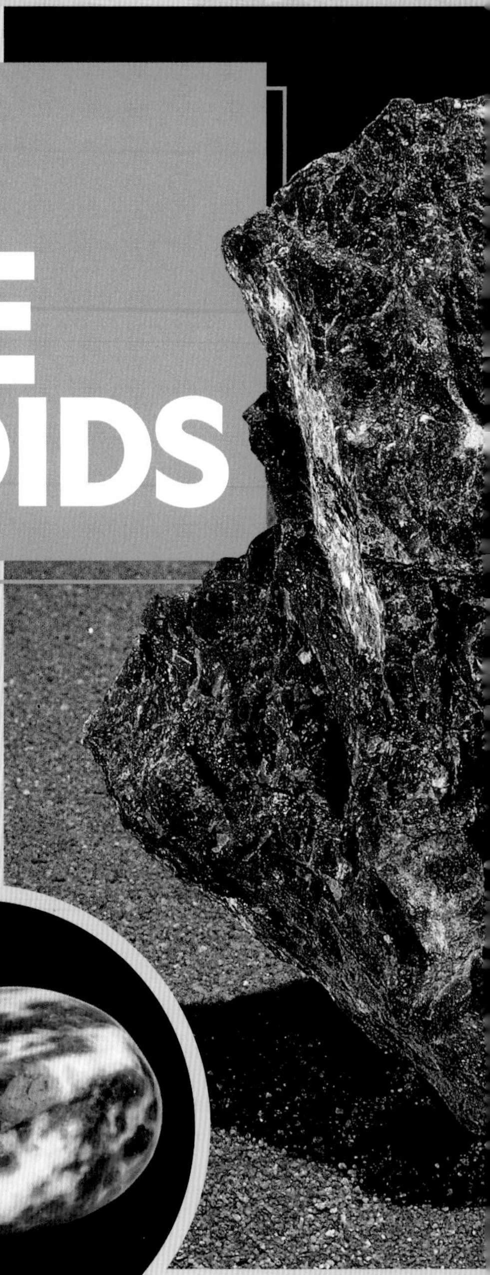

Suppose that you were hungry and decided to whip up a peanut butter and jelly sandwich, but then you discovered that you are out of jelly. What would you do? Well, if you don't like eating plain peanut butter, you would use something else that's sweet in place of the jelly—like honey, perhaps. This is similar to what's going on with the feldspathoid minerals. As the name suggests, they are similar to feldspars (pp. 212–213) and form in the same types of igneous rocks. The problem is, the magma they form in does not have enough silica to form true feldspars, so other chemicals fill in their structure as they crystallize. This gives the feldspathoid minerals a flavor all their own. Here are five varieties of fantastic feldspathoids!

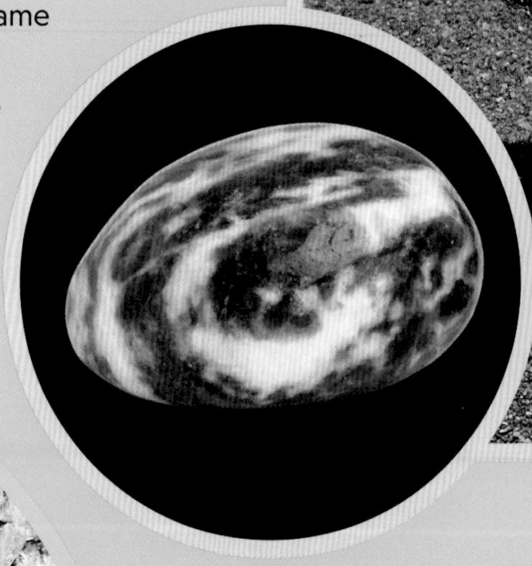

NEPHELINE

Nepheline, the most common of the feldspathoid minerals, is usually found in both intrusive and extrusive igneous rocks, especially pegmatites. It comes from mafic magmas that do not have a lot of silica in them. It rarely forms well-developed crystals, but when it does, the crystals are shaped like six-sided prisms, they are colorless or white, and they have a glassy luster. Most often, nepheline forms in massive deposits with no visible crystals. Nepheline is sometimes used in place of feldspars in making ceramics.

CANCRINITE

Cancrinite is a relatively rare feldspathoid mineral that often forms when nepheline and feldspars found in mafic igneous rocks change over time. It can also be found in the area around the surface of limestones (pp. 120–121) that have become metamorphosed when magma flows up through cracks in the rock. Cancrinite comes in a variety of colors including blue, gray, and green, but it is typically white. It usually forms in large masses, but sometimes cancrinite will form well-developed prism-shaped crystals that have a glassy luster.

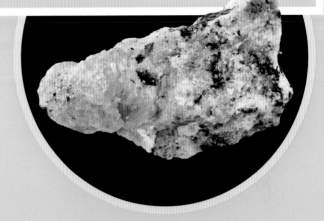

SODALITE

Sodalite is often confused with lazurite because both have a deep blue color, and both can be found in limestones that have undergone contact metamorphism. Most often sodalite is found along with the mineral nepheline in mafic igneous rocks such as phonolite. Sodalite can also be shades of gray or green and is mostly found in a massive form, but sodalite can form crystals which have a glassy luster. One of the best places to find sodalite crystals is in the lava flows on Mount Vesuvius, a famous volcano in Italy. Sodalite is sometimes cut and polished to make beads for jewelry. Its name comes from the fact that it is rich in the chemical element sodium.

LAZURITE

Lazurite is a famous feldspathoid because along with the minerals sodalite, pyrite (pp. 160–161), and calcite (pp. 190–191) it is one of the main components forming the gemstone lapis lazuli. Lazurite is known for its deep blue or sometimes greenish blue color and is usually found in limestones (pp. 120–121) that have undergone contact metamorphism. Lazurite is most often found in a massive granular form without visible crystals. When crystals do form, they are generally shaped a little like soccer balls, and they can have a dull to glassy shine. In the past, lapis lazuli was ground up to make a blue pigment called ultramarine, but today it is usually cut and polished to be used as a decorative stone or in jewelry.

LEUCITE

Leucite is different from many of the other feldspathoid minerals. While it does form in massive deposits, it is also commonly found as crystals growing in spaces in basalt (p. 90) lava flows that are rich in the mineral olivine (pp. 234–235). The crystals often have a very odd shape called a trapezohedron, which looks a little like a sphere made of lots of flat panels. Because of their white or light gray color, the crystals usually stand out against the dark lava. In fact, the mineral's name comes from the ancient Greek word *leukos,* which means "white." Leucite is rich in the element potassium and was often ground up and used as a fertilizer for plants.

MARVELOUS
MICAS!

Known as phyllosilicates, many of the minerals in this group have a fun property that people love to explore. Like all silicate minerals, the phyllosilicates are made from a pyramid-shaped building block called the silicon tetrahedron. But in these minerals, the silicon tetrahedra are arranged in large flat sheets stacked on top of each other. You can literally peel them apart with your fingers! Mica minerals are extremely important in the formation of both igneous and metamorphic rocks and are a major component of sedimentary rocks, too. Here are five marvelous mica minerals to check out.

LEPIDOLITE

Lepidolite is one of the more colorful members of the mica group. It often comes in shades of violet, lilac, and yellow, along with the gray and colorless varieties. This unique coloration is due to the fact that lepidolite can also contain the element manganese along with lithium. Lepidolite is actually one of the largest sources of lithium, an important metal. Lepidolite is often found in bunches of scaly-looking, six-sided, plate-shaped crystals, but it can also come in a massive botryoidal form, which looks like a bunch of grapes. It is most commonly found in granite pegmatite along with beryl (pp. 230–231) and topaz (pp. 238–239). Lepidolite is named for the way its crystals form. The Greek term *lepidos* means "scale."

PHLOGOPITE

Phlogopite is a yellow to reddish brown mica mineral that is rich in magnesium. It is part of the group of micas that includes the darker-colored biotite. But unlike biotite, phlogopite is not found in granite (pp. 78–79) where quartz (pp. 206–207) is common. It is more at home in ultramafic rocks such as a kimberlite (pp. 154–155). Along with calcite (pp. 190–191), it is most commonly found in metamorphic marble (pp. 100–101), where it forms six-sided, prism-shaped crystals and platelike scales. Like other micas, it can be peeled into thin sheets that reveal a starlike pattern when light passes through them. Because of its coloring, phlogopite gets its name from the Greek word *phlogopos*, which means "fiery looking."

MUSCOVITE

Most people picture the mineral muscovite when they hear the term "mica." That's why muscovite is also called common mica. Muscovite is a main component in felsic igneous rocks such as granite (pp. 78–79) and in metamorphic schists (p. 107) and gneiss (p. 106). It can often be found in veins deposited in other types of rocks by hot, mineral-rich water that has flowed up through cracks. Like all micas, muscovite forms large, flat, platelike crystals that are usually clear, white, or silvery gray. Often the crystals form a thick layered structure called books. And some of these books are huge, measuring almost 10 feet (3 m) across! Of course, muscovite's most interesting property is its perfect cleavage, which allows you to peel off large sheets. In fact, this property is how muscovite got its name. In the past, large sheets of this mineral were used as windows by people of the Muscovy province in Russia, who called it Muscovy glass.

CHLORITE

Technically speaking, "chlorite" is not the name of a single mineral. It is a term used to describe a group of green-colored mica minerals that have the same chemical and physical properties. Chlorite minerals are rich in iron and magnesium, and they tend to form six-sided or prism-shaped crystals. They can also sometimes make thick masses of flakes. Normally dark green to almost black in color, chlorite minerals often form from other iron-rich minerals such as pyroxenes (pp. 224–225) and amphiboles (pp. 228–229) when they undergo low-grade metamorphism or have hot, mineral-rich fluids flow through them. Because of its typical green color, this mineral group gets its name from the Greek term *chloros*, which means "green."

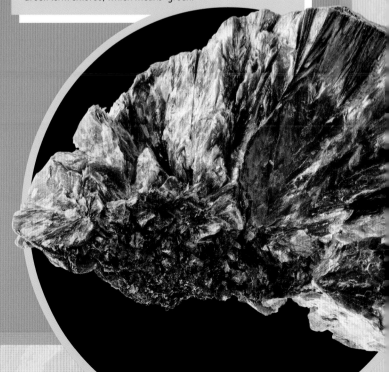

BIOTITE

Biotite is the name that is used to describe a series of dark-colored mica minerals containing iron and/or magnesium. It is usually black, brown, yellow, or even bronze. Like muscovite, biotite is a major component of metamorphic schists (p. 107) and gneiss (p. 106), as well as the igneous rocks granite (pp. 78–79), diorite (pp. 80–81), and even some gabbro (pp. 82–83). Most often biotite will either form flat platelike crystals or short barrel-shaped prisms, but these crystals tend to be much smaller than those formed by muscovite. Like all micas, biotite has a perfect cleavage that allows you to split it into sheets. And even though it is slightly harder than muscovite, it is still soft enough that you can scratch it with the edge of a penny.

Serpentine is the name of a group of similar silicate minerals that are often mixed together to produce rocks with very different appearances.

Bet You Didn't Know!

In the past, the chrysotile form of serpentine was mined to make "white asbestos," which was used as a fireproofing material that was sprayed inside buildings. Eventually, it was discovered that little pieces of the fibers could break off. When people breathed them in, it caused all sorts of serious medical conditions. As a result, asbestos is no longer used in new buildings. When it is found in older buildings, it is usually removed.

THE SERPENTINE GROUP

Imagine that you're walking along a forest path when all of a sudden, you notice a long, dark green object sticking out from under some ferns. It has a kind of mottled surface with darker patches mixed in with the green, and it looks an awful lot like the body of a big snake. You're all set to run for help when you see a chipmunk scurrying into a hole right under the "snake." It turns out that your snake is really an outcrop of serpentine. Don't feel too bad—this mineral gets its name from the fact that it looks like snakeskin!

In the past, serpentine was thought to be a single mineral. Now geologists recognize that it is actually a group of very closely related minerals that are often mixed together to the point that it's hard to tell them apart. Like the mica (pp. 216–217) minerals, the minerals in the serpentine group have a layered structure that can sometimes produce long fibers. All of the minerals in the group are relatively soft, and they are found in mafic igneous rocks such as peridotite that have undergone metamorphism due to hot, mineral-rich water flowing through them. Most often, these minerals form as a result of other silicate minerals such as pyroxenes (pp. 224–225), olivine (pp. 234–235), and amphiboles (pp. 228–229) being changed over time.

Geologists have identified more than a dozen separate minerals in the serpentine group. The two most common varieties of serpentine are called antigorite and chrysotile. Antigorite does not usually form large crystals. Instead, it forms in masses with overlapping plates that can be split apart almost like mica. Chrysotile is usually a lighter greenish white color and has a fibrous surface. It's best to never handle serpentine minerals. Even though you won't get a snakebite from this serpent-skinned rock, the fibers can be very harmful if they break off and you accidentally breathe them in.

FACTS

CLASS: silicate minerals—phyllosilicates

MAIN COLORS: green, gray-green, yellow, white

CLEAVAGE: present but not visible; conchoidal to splintery fracture

LUSTER: greasy, resinous, earthy

STREAK: white

HARDNESS: 2.5–5, depending on specific mineral species

OTHER DISTINCTIVE FEATURES: often fibrous, dark green color with greasy feel

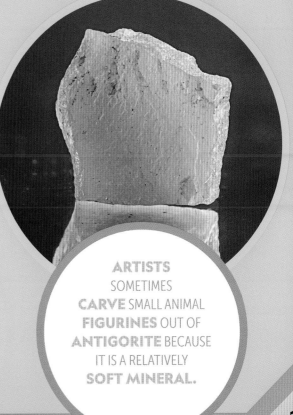

ARTISTS SOMETIMES CARVE SMALL ANIMAL FIGURINES OUT OF ANTIGORITE BECAUSE IT IS A RELATIVELY SOFT MINERAL.

TALC

When Mr. Mohs was creating his famous scale of mineral hardness, he needed to find a mineral that was softer than all the rest to serve as the anchor. After much testing, he found one that could be scratched by every other mineral and could scratch no other. That mineral was talc, numero uno on Mohs scale and a mineral that is so soft, you can break it apart just by rubbing it between your fingers!

Talc is considered to be a secondary mineral that forms when rocks like dolostone (pp. 194–195) and peridotite become heated and begin to change into a metamorphic rock. It is especially common to find talc where the serpentine (pp. 218–219) minerals are found, and it can also be found in some schists (p. 107). During the metamorphic process, minerals that have lots of magnesium in them, such as olivine (pp. 234–235), pyroxenes (pp. 224–225), and amphiboles (pp. 228–229), change into talc. In many cases, the talc will keep the shape of the original mineral's crystals.

Talc gets its name from the Arabic word *talq,* which means "mica." Even though it does not peel off in sheets like pieces of muscovite (p. 217) do, talc does have the same type of cleavage as mica minerals and belongs in the same mineral class. On its own, talc rarely forms well-developed crystals. Instead, it forms foliated masses that sometimes look like thin plates stacked on top of each other. Talc also forms large, massive layers in the rock. When this happens, the rock is called soapstone (pp. 110–111). Soapstone is an important resource because it is so soft that it can easily be cut and shaped into things like countertops and tables. Artists also use it to make all sorts of sculptures. When it comes to being soft, talc is the big winner!

FACTS

CLASS: silicate minerals—phyllosilicates

MAIN COLORS: colorless, white, pale apple green to dark green

CLEAVAGE: perfect, uneven fracture; sometimes conchoidal

LUSTER: pearly to greasy

STREAK: white

HARDNESS: 1

OTHER DISTINCTIVE FEATURES: greasy feel

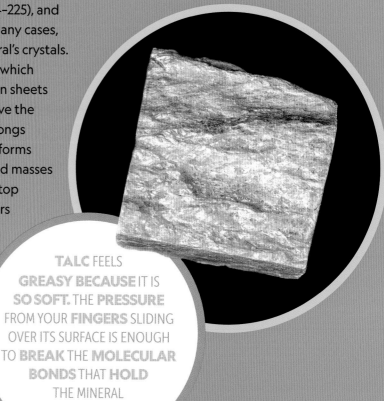

TALC FEELS GREASY BECAUSE IT IS SO SOFT. THE PRESSURE FROM YOUR FINGERS SLIDING OVER ITS SURFACE IS ENOUGH TO BREAK THE MOLECULAR BONDS THAT HOLD THE MINERAL TOGETHER.

Talc is usually an apple green color and shines with greasy or pearly luster.

Bet You Didn't Know!

Because talc is so soft and easily breaks into small plate-shaped particles, it has many industrial uses and can be found in many of the products that we use in our daily lives. Talc is often used as a filler material in plastics to make them stiffer and more resistant to heat. It is also added to paint to help keep it from separating after it has been mixed. Talc is also often added to writing paper to make it smoother and shinier.

BORN TO BE BEAUTIFUL

People have long believed that certain gems and crystals have special powers to bring good health, good luck, or just make you happy. Nobody knows for sure, but this is probably where the idea of birthstones came from. Basically, a birthstone is a specific gem that goes along with the month that a person was born in. The idea of birthstones dates back to a time before the modern calendar, and over the years some of the stones have changed. Today's modern list of birthstones was updated by jewelers to include only transparent gems that make it easier to produce pieces of jewelry. Most of the birthstones on the original list are still the same though. Below is the updated list with the name of the traditional birthstone in parentheses.

ABOUT THE **FIRST CENTURY** A.D., PEOPLE BEGAN **CONNECTING SPECIFIC STONES** WITH EACH OF THE **12 SIGNS** OF THE **ZODIAC.** OVER TIME, THESE BECAME THE **BIRTHSTONES** THAT WE HAVE **TODAY.**

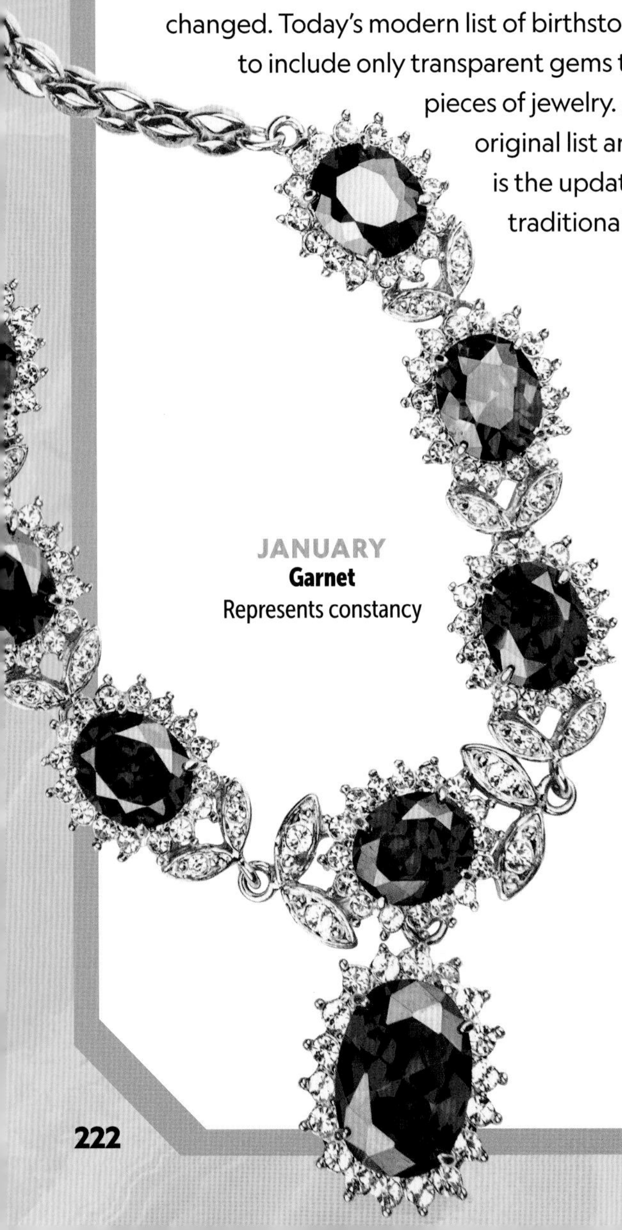

JANUARY
Garnet
Represents constancy

FEBRUARY
Amethyst
Represents sincerity

MARCH
Aquamarine
(originally bloodstone)
Represents courage

222

MAY
Emerald
Represents love
and success

APRIL
Diamond
Represents innocence

JUNE
Alexandrite (originally pearl)
Represents health and longevity

JULY
Ruby
Represents
contentment

SEPTEMBER
Sapphire
Represents
clear thinking

AUGUST
Peridot (originally sardonyx)
Represents happiness in marriage

DECEMBER
**Blue zircon
(originally turquoise)**
Represents prosperity

OCTOBER
**Pink or green tourmaline
(originally opal)**
Represents hope

NOVEMBER
**Citrine
(originally topaz)**
Represents fidelity

POWERFUL PYROXENES!

As a group, members of the pyroxene family of minerals are some of the most important rock-forming silicate minerals both on Earth and in space. Found in many types of igneous and metamorphic rocks in Earth's crust, these dark-colored minerals are believed to play a major role in forming the rocks of the mantle. Some pyroxenes have even been discovered in meteorites and moon rocks. Pyroxenes belong to a group of minerals called inosilicates. Like all silicate minerals, pyroxenes are made from building blocks called silicon tetrahedra. What makes inosilicates different is that their silicon tetrahedra form long chains. This structure gives pyroxene minerals a well-developed cleavage that often looks like a set of steps on the side of the crystals. Here's a look at five pyroxenes that really pack a punch, geologically speaking.

DIOPSIDE

Diopside crystals are usually some shade of green (often bright green) but can also be colorless or white. Diopside is most commonly found in regional metamorphic rocks—especially marbles (pp. 100–101)—that have formed from limestone (pp. 120–121) and dolostone (pp. 194–195) and have lots of quartz sand in them. It is also found in contact metamorphic rocks that are rich in iron. Diopside forms short prism-shaped crystals that sometimes have a layered look to them. Diopside crystals often have two sets of prism faces that face in different directions. That feature is how diopside got its name. In Greek, *di* means "two" and *opsis* means "vision," but it is most often translated as "double appearance."

AUGITE

Augite, the most common of the pyroxene minerals, is part of a grouping of pyroxene minerals that includes diopside. Augite often forms short but well-developed prism-shaped crystals in mafic igneous rocks, such as gabbro (pp. 82–83) and diorite (pp. 80–81), and grainy masses in basalt (p. 90). It can also be found in rocks that have been metamorphosed at high temperatures. Augite is usually dark green, brown, or black and often has a glassy luster. This explains its name. The word "augite" is believed to have come from the Greek word *augites*, which means "brightness." In addition to being common in many rocks here on Earth, augite is also a main component in both stony meteorites and basaltic rocks that were brought back from the moon. It is one mineral that is truly out of this world!

ENSTATITE

Enstatite, a pyroxene that is rich in the chemical element magnesium, is commonly found in mafic igneous rocks such as gabbro (pp. 82–83) and peridotite. It has also been found in both iron and stony-iron meteorites, so we know that it can form in space, too. Enstatite is part of a grouping of pyroxenes, including hypersthene, that have similar physical properties. While enstatite can form prism-shaped crystals, it usually just occurs as masses that are gray, dark yellow, or greenish brown in color.

SPODUMENE

While the color of most pyroxene minerals tends to be on the dark side, spodumene is often bright and colorful. Spodumene is named for its common white or light ash gray color and comes from the Greek term *spodoumenos,* which means "burnt to ash." But, unlike other pyroxenes, spodumene can also be bright green, purple, or pink because it contains the chemical elements aluminum and lithium in its structure, which produce different colors. Like lepidolite, spodumene is an important source of lithium. Unlike most other pyroxenes that form in mafic igneous rocks, spodumene forms only in granite pegmatites, where it produces long, flattened crystals with deep grooves on the surface. Some spodumene crystals can be quite large. The heaviest on record comes from the Etta Mine in South Dakota, U.S.A. This whopper measured 42 feet (13 m) long and weighed in at well over 50 tons (45 t). That's as much as some train engines!

HYPERSTHENE

Hypersthene belongs in the same mineral grouping as enstatite and it has traditionally been classified as a separate mineral. However, a detailed chemical analysis of the two minerals determined that hypersthene is really just an iron-rich form of the mineral enstatite. Hypersthene is usually a dark green, gray, or brown with more of a bronze luster than "regular" enstatite, and well-developed crystals are rare. Instead, hypersthene is commonly found as coarse grains in both gabbro (pp. 82–83) and peridotite, where it is often confused with the amphibole mineral hornblende (p. 228). The easiest way to tell them apart is to look at the cleavage. In hypersthene, the edges of the mineral grains look like they have little steps going up at right angles. In hornblende, the edges are sloped.

Jade is a precious gemstone that comes from the mineral jadeite and is often used to make spectacular carvings.

WERAWSANA JADE PAGODA, MYANMAR (BURMA)

Bet You Didn't Know!

It is extremely difficult to tell jade that comes from jadeite apart from the jade that comes from the amphibole mineral nephrite. They are so close in appearance that it wasn't until 1863 that a French mineralogist named Alexis Damour realized that they actually come from two different minerals.

JADEITE

When it comes to precious gemstones, few can top jade. While it may not form beautiful crystals like emeralds (pp. 230–231) and diamonds (pp. 152–153), its mesmerizing green color has made it a favorite among people dating back to the Stone Age. Jade is the term used to describe a hard, massive, green rock that can be polished to a glassy luster, and it comes from two different silicate minerals. Lower-quality jade comes from the amphibole mineral nephrite (p. 229), but the highest quality jade, known as imperial jade, only comes from the pyroxene mineral jadeite.

Jadeite is often found along with serpentine in metamorphic rocks that started with large amounts of sodium feldspars in them. It almost never forms crystals that you can see. Usually, the mineral is made up of interlocking fibers or blades. It can also form when hot water rich in sodium flows through other rocks, replacing the original minerals. Because it is very resistant to weathering, jadeite does not break down when the rocks around it wear away. Instead, it is carried by the flowing water and gets concentrated in low spots at the bottom of the streams, forming placer deposits.

While known mostly for its green color, jadeite can also come in shades of white with green spots. Sometimes it can even be a light purple. When found in the field, jadeite has either a dull or waxy luster, but when it is polished it shines like glass!

People first used jade during the Stone Age as cutting tools. Later, as metals came into use, it became more popular for making jewelry and sculptures. While ancient jade artifacts have been found all over the world, some of the finest pieces come from China, where people have been working with jade for thousands of years. It was also very popular in what is now Mexico and Central America, where people of the Aztec and Maya cultures used it to make ornaments. Even today, people love to wear jewelry made from jadeite jade.

FACTS

CLASS: silicate minerals—inosilicate

MAIN COLORS: apple green, emerald green, white with green spots, and, rarely, purple

CLEAVAGE: distinct in two directions

LUSTER: glassy when polished, waxy or even dull in field

STREAK: white

HARDNESS: 6–7

OTHER DISTINCTIVE FEATURES: forms hard, tough masses

JADEITE CAN BE FOUND **ALL OVER THE WORLD,** BUT SOME OF THE **LARGEST DEPOSITS** ARE LOCATED IN THE COUNTRY OF **MYANMAR (BURMA).**

AWESOME AMPHIBOLES!

Like pyroxenes, the family of minerals known as amphiboles are inosilicates. That means they were formed by long chains of silicon tetrahedra, the chemical compound that looks like a tiny pyramid and is the main building block of all silicate minerals. While pyroxenes are built with a single chain of tetrahedra, amphiboles have a double chain. These two families of minerals are often confused for each other because they form in the same rock types. The easiest way to tell them apart is to examine how they break along their edges. When you look at amphibole minerals from the side, the cleavage is slanted. Pyroxenes have a square-edged, steplike cleavage. Here are five totally awesome amphiboles that form the backbone of many igneous and metamorphic rocks.

HORNBLENDE

Hornblende isn't a single mineral, but rather it's a name that geologists use to describe a group of closely related amphibole minerals that have pretty much the same properties. The only major difference between them is their chemical composition. Hornblende minerals are usually the most commonly identified amphibole in metamorphic rocks including gneiss (p. 106), schist (p. 107), and especially amphibolite (pp. 104–105). They are also a major component in mafic igneous rocks such as diorite (pp. 80–81). Hornblende is usually dark green, brown, or black, and it forms glassy prism-shaped crystals with a diamond-shaped cross section.

GLAUCOPHANE

Glaucophane is an important mineral for geologists. When they find it in metamorphic rocks, it gives them an idea of what the temperatures and pressures were when the rocks formed. Glaucophane, which has a characteristic bluish black or lavender blue color, is found in schist (p. 107) and marble (pp. 100–101) that were formed under high-pressure but low-temperature regional metamorphism. These types of rocks are sometimes called blueschists and form in areas where an oceanic tectonic plate is sliding under a continental plate. Glaucophane is usually found with jadeite (pp. 226–227), epidote, and garnet (pp. 242–243), and forms thin crystals that can be either blades or prism-shaped.

TREMOLITE

Tremolite is a light-colored mineral that ranges from white to dark gray but can also be yellow or pink. It is part of a chemical grouping with another amphibole called actinolite. Tremolite is rich in the element calcium and most commonly forms when limestone (pp. 120–121) and dolostone (pp. 194–195) undergo low- to medium-grade metamorphism. It is also found in schist (p. 107) along with the mineral talc (pp. 220–221). Tremolite forms several interesting varieties of crystals. Sometimes they appear as long blades that fan out from a central point. Other times the crystals look feathery or form long fibers.

ACTINOLITE

Actinolite is an iron-rich, dark-colored amphibole that is usually green, grayish green, or black. It is part of a chemical grouping with the mineral tremolite. Most often, it forms long prism-shaped crystals with a diamond-shaped cross section that can sometimes fan out from a central point. Because of this trait, it gets its name from a Greek term that means "ray." Actinolite is very common in low- to medium-grade metamorphic rocks, especially schist (p. 107).

NEPHRITE

Nephrite is the name given to an amphibole mineral that is actually a type of actinolite or tremolite. But instead of forming well-developed crystals like actinolite and tremolite do, nephrite is found as dense, hard, compact masses of interlocking fibers. Since its form is so different from that of its two "parents," nephrite is often treated as a separate mineral. Nephrite can come in a wide variety of colors. When it contains lots of iron and has the composition of actinolite, it's dark green. When it has the composition of tremolite and has lots of calcium in it, nephrite is creamy white. It can also be gray, brown, or even lavender. Nephrite is commonly found along with talc (pp. 220–221) and serpentine (pp. 218–219) in metamorphic rocks. Nephrite and the pyroxene mineral jadeite (pp. 226–227) are the two minerals that produce the gemstone jade.

BERYL

You may not be familiar with the mineral name "beryl," but you have probably heard some of the other names that this mineral goes by when it takes the form of a gemstone. It can be blue-green aquamarine, pink morganite, yellow-brown heliodor, and the king of them all, emerald, which is bright "emerald" green. Common, ordinary beryl is usually white or pale green, but it can also be blue, red, and just plain colorless and clear (a variety called goshenite).

Beryl is actually a fairly common silicate mineral that is found in many felsic igneous rocks such as rhyolite (p. 90) and especially granite pegmatite. These rocks usually include the minerals quartz (pp. 206–207), biotite mica (pp. 216–217), and microcline feldspar (pp. 212–213). It is also found in garnet (pp. 242–243) and quartz in schists (p. 107) that have undergone regional metamorphism.

Beryl is usually fairly easy to identify because it has some rather unique physical properties. It is a superhard mineral, rating between 7.5 and 8 on the Mohs scale. It also forms distinctive crystals in the shape of six-sided hexagonal prisms, which are sometimes confused for quartz—until you take a closer look. Beryl crystals almost always have long thin lines called striations running down the long part of the crystal. Quartz can also have striations, but when it does, the lines run across the crystals.

Beryl belongs to a group of minerals called cyclosilicates. Like all silicate minerals, cyclosilicates are made from chemical building blocks, called silicon tetrahedra, that look like little pyramids. The silicon tetrahedra in cyclosilicates are arranged in rings, unlike in amphiboles and proxenes, which have silicon tetrahedra in chains. In the case of beryl, the ring is made from six tetrahedra. This is one of the reasons that beryl crystals have a six-sided shape.

FACTS

CLASS: silicate minerals—cyclosilicates

MAIN COLORS: white, bright green (emerald), blue, greenish blue (aquamarine), yellow, red, pink (morganite)

CLEAVAGE: poor in one direction, uneven fracture sometimes conchoidal

LUSTER: glassy

STREAK: none

HARDNESS: 7.5–8

OTHER DISTINCTIVE FEATURES: striations along the length of the crystals

BERYLLIUM

BERYL IS A GOOD SOURCE OF THE METAL BERYLLIUM, WHICH IS MIXED WITH OTHER METALS (ESPECIALLY COPPER) TO MAKE ALLOYS—SOME OF WHICH ARE USED IN ELECTRONIC EQUIPMENT SUCH AS COMPUTERS, CELL PHONES, AND MEDICAL DEVICES.

Beryl comes in a rainbow of colors and is used for the gemstones aquamarine (blue-green) and emerald (green).

Bet You Didn't Know!

Perfect emeralds are quite rare and most have tiny flaws called inclusions in them. One of the things that makes emeralds so popular as a gem is the way their green color, which comes from small amounts of the chemical element chromium, is evenly spread throughout the entire crystal.

231

Tourmaline is a silicate mineral that comes in a rainbow of colors, sometimes all in a single crystal!

Bet You Didn't Know!

"Watermelon tourmaline" is a variety of the tourmaline called elbaite that can have a dark green "rind" around the outside of the crystal surrounding a dark pink center. It looks just like a watermelon that has been sliced in half!

TOURMALINE FAMILY

Do you like rainbows? Most people do! When the sun shines right after a storm passes, you can often see a bright band of colors stretched across the sky. Now imagine holding a rainbow in the palm of your hand. While this sounds like a myth, it is possible if you have the right chunk of the mineral tourmaline.

The name "tourmaline" is used for an entire group of more than a dozen main minerals (and a bunch of less common varieties) that have the same basic crystal structure but have a wide range of chemical compositions and colors. Like beryl (pp. 230–231), tourmaline is a member of the cyclosilicate class of minerals, which means that it is made of chemical building blocks called silicon tetrahedra that look like tiny pyramids. These pyramids are arranged in rings, but between the rings are spaces that can have lots of different chemical elements in them. It's these chemical differences that create the different minerals and the wide range of colors.

Tourmaline is actually a very common mineral, but different varieties are found in different rock types. Two of the most common types are colorful elbaite and black schorl, which are both found mainly in granite pegmatites. Another variety called dravite forms in marble (pp. 100–101), and buergerite can be found in rhyolite (p. 90).

No matter what the color or the source rock is, all varieties of tourmaline tend to form long prism-shaped crystals that have a triangle-shaped cross section. Sometimes the crystals can also be short and stubby, but they usually show distinct lines called striations running down the length of the crystal. Because it is a relatively hard mineral, sometimes you can find tourmaline crystals that have eroded from their source rocks mixed in with gravel in streams.

FACTS

CLASS: silicate minerals—cyclosilicates

MAIN COLORS: black, brown, green, yellow, pink, blue, multicolored

CLEAVAGE: none; uneven fracture sometimes conchoidal

LUSTER: glassy

STREAK: white

HARDNESS: 7–7.5

OTHER DISTINCTIVE FEATURES: crystals have a rounded triangular cross section

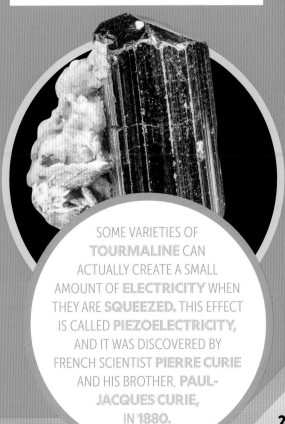

SOME VARIETIES OF **TOURMALINE** CAN ACTUALLY CREATE A SMALL AMOUNT OF **ELECTRICITY** WHEN THEY ARE **SQUEEZED.** THIS EFFECT IS CALLED **PIEZOELECTRICITY,** AND IT WAS DISCOVERED BY FRENCH SCIENTIST **PIERRE CURIE** AND HIS BROTHER, **PAUL-JACQUES CURIE,** IN **1880.**

OLIVINE

Sometimes the name of a mineral can tell you a lot about its properties. Take olivine, for example. The name "olivine" has to do with the fact that its color is very similar to that of ripe green olives.

The term "olivine" describes a group of minerals that share the same crystal structure and physical properties but have slightly different chemical compositions. Olivine minerals belong to the simplest type of silicate minerals called nesosilicates. Like all silicate minerals, they are made from individual chemical building blocks called silicon tetrahedra that are shaped like little pyramids. Instead of being linked together as they are in other silicates, the tetrahedra in olivine bond directly to other chemical elements. Sometimes they bond to magnesium, forming a type of olivine called forsterite. Other times they bond to iron, making an olivine called fayalite. There are also varieties of olivine created from a mix of both iron and magnesium.

Rarely does olivine form well-developed crystals. Instead, it is usually found as rounded grains or simply as a solid mass. Olivine is a very important rock-forming mineral and can be found in many types of mafic igneous rocks that are rich in iron and magnesium. Some of these include basalt (p. 90), gabbro (pp. 82–83), and peridotite. Olivine is also believed to make up much of the rocks found in Earth's upper mantle. If that weren't enough, it has even been found in some meteorites from space!

FACTS

CLASS: silicate minerals—nesosilicate

MAIN COLORS: olive green, yellowish green, yellowish brown, may be transparent

CLEAVAGE: poor in two directions, uneven fracture, sometimes conchoidal

LUSTER: glassy

STREAK: colorless

HARDNESS: 6.5–7

OTHER DISTINCTIVE FEATURES: yellowish green color and rounded grains

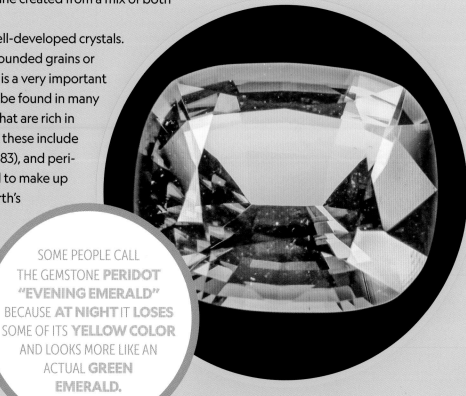

SOME PEOPLE CALL THE GEMSTONE **PERIDOT** "EVENING EMERALD" BECAUSE **AT NIGHT** IT **LOSES** SOME OF ITS **YELLOW COLOR** AND LOOKS MORE LIKE AN ACTUAL **GREEN EMERALD.**

Olivine only rarely forms well-developed crystals, but it is frequently found as small crystal grains in many types of igneous rocks.

Bet You Didn't Know!

When olivine forms large, yellow-green transparent crystals it is used as a gemstone called peridot. Unlike other gemstones, when peridot crystals form, they have few inclusions, or flaws, making it easy for jewelers to cut the stones. People have been mining peridot for thousands of years, and it is currently the birthstone for August.

Zircons can often form large crystals in both igneous and metamorphic rocks.

Bet You Didn't Know!

Sometimes jewelers will "enhance" the color of a zircon by heating the stones to a high temperature. Using this process, common brown zircons can be turned into blue ones, which are much more popular. Over time, however, the colors of the heat-treated stones can fade, especially when they are left in bright sunlight.

ZIRCON

Zircon is a pretty cool name for a mineral, although it sounds more like the name of a distant planet. The truth is, the name "zircon" comes from the ancient Persian term *azargun,* which basically means "gold-colored." This is a bit deceiving because zircons aren't usually gold. Most zircons tend to be dark brown, black, gray, yellow, green, orange, red, and even clear.

Like olivine (pp. 234–235), zircon is a nesosilicate mineral, which means that it has a simple structure made from a single silicon tetrahedron attached to an atom of the chemical element zirconium. Zircon often forms well-developed crystals in the form of prisms that have a square cross section and pointed pyramids at each end. When they are short, they look like an eight-sided octahedron. In some cases, individual zircon crystals can get quite large, with some weighing in at over eight pounds (3.6 kg).

Zircons are commonly found in small amounts in all types of igneous rocks but occur in greater amounts along with orthoclase feldspar (pp. 212–213) and biotite in igneous felsic pegmatite rocks. They are also found in a wide range of metamorphic rocks including gneiss (p. 106), schist (p. 107), and even some marble (pp. 100–101). Some of the best places to find zircon crystals are in streams and at beaches where flowing waters have concentrated them with gravel after they have weathered out of the rocks where they formed.

While zircon is mined as a source of the metal zirconium, certain varieties have been used as gemstones for thousands of years, especially when they form transparent crystals. Some of the more common varieties include hyacinth zircon, which can be red, orange, or yellow, and starlite zircon, which is blue. It turns out that when they are properly cut, zircons can shine with a better brilliance than even diamonds.

FACTS

CLASS: silicate minerals—nesosilicate

MAIN COLORS: very variable, including dark brown, black, gray, yellow, green, orange, red, clear

CLEAVAGE: poor in two directions with uneven fracture

LUSTER: glassy

STREAK: colorless

HARDNESS: 7.5

OTHER DISTINCTIVE FEATURES: denser than most common minerals

ZIRCON GEMSTONES SHOULD NOT BE CONFUSED WITH **CUBIC ZIRCONIA,** WHICH ARE **ARTIFICIAL GEMS** MADE FROM THE **ELEMENT ZIRCONIUM** TO **LOOK LIKE A DIAMOND.**

TOPAZ

Topaz is a rare mineral in more ways than one. First, even though it is a popular gemstone that people have collected for hundreds of years, finding "clean" transparent crystals is not all that easy. When topaz forms crystals they often have inclusions, so very few stones are considered to be 100 percent flawless.

The second rarity has to do with how topaz forms. Topaz is most often found in felsic igneous rocks in veins of granite pegmatite and in spaces found in rhyolite (p. 90) lava flows. Unlike other igneous minerals, topaz crystals do not grow directly from hot liquid magma as it cools. Instead, topaz crystals grow directly from superhot gases that contain large amounts of the element fluorine. These gases get released from magma or lava after most of the other minerals have already formed, so topaz is one of the last minerals to get deposited.

Topaz comes in a wide range of colors, including white, yellow, brown, orange, pink, and gray. It can also be colorless. Its crystals can be either translucent or completely transparent. The crystals that form are usually prism-shaped but the ends can be pointed in different directions. In most cases, the crystal faces have lines called striations in them that run along the length of the crystal. Some topaz crystals can get to be very large. In fact, several really big topaz crystals were found in Brazil, with the largest weighing almost 600 pounds (270 kg). That's about the same weight as three adult men!

Topaz, the original birthstone for the month of November, is a very popular gemstone. The most valuable type of stone is called imperial topaz, which has an orange to orange-pink color. Other popular colors are pink, purple, and deep blue, but these colors are rarely found in natural stones. Treating clear, brown, or yellow topaz with radiation and heat can produce these other colors, but they can sometimes fade over time.

FACTS

CLASS: silicate minerals—nesosilicate

MAIN COLORS: white, colorless, yellow, brown, orange, pink, gray; can also have a blue or green tint

CLEAVAGE: perfect in one direction; uneven fracture sometimes conchoidal

LUSTER: glassy; can be transparent or translucent

STREAK: colorless

HARDNESS: 8

OTHER DISTINCTIVE FEATURES: extremely hard and splits along cleavage plane

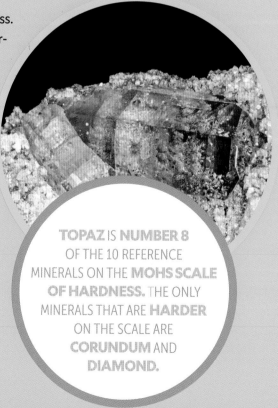

TOPAZ IS **NUMBER 8** OF THE 10 REFERENCE MINERALS ON THE **MOHS SCALE OF HARDNESS.** THE ONLY MINERALS THAT ARE **HARDER** ON THE SCALE ARE **CORUNDUM** AND **DIAMOND.**

A child sits with two uncut topaz crystals and a large faceted topaz.

Topaz is a popular gemstone that forms in an unusual way.

Bet You Didn't Know!

Even though topaz is very hard, it's not very strong. The difference is often confusing. Hardness is a measure of how easy it is to scratch a mineral. The property that measures how easy a mineral is to break is called tenacity, and on that scale, topaz is listed as very brittle. Because of this, a topaz gem in a ring can easily split if the stone is hit the wrong way.

MAGICAL
METAMORPHIC
MINERALS!

Even though geologists don't usually believe in magic, many will agree that something magical happens when a rock gets transformed by heat and pressure and eventually becomes an entirely new metamorphic rock. While some minerals like quartz (pp. 206–207) and feldspar (pp. 212–213) can show up in all three major rock types, there are some that mostly form in a metamorphic environment. These "magical" minerals can be really helpful to geologists because knowing how they form gives us clues about the geological history of an area.

KYANITE

Kyanite often produces long, flat, tabular or bladed crystals. It can come in shades of green or gray, but kyanite is famous for its deep blue color. One of the strange things about kyanite crystals is that the hardness of the mineral is different depending on the direction you test it. When you scratch it along the length of the crystal it can be as soft as 4 on the Mohs scale. But when you scratch it across the crystal, the hardness jumps up to a 6 or 7. Like andalusite and sillimanite, kyanite is found in gneisses (p. 106) and schists (p. 107) that have undergone regional metamorphism. But in the case of this mineral, the temperature was fairly low while the pressure was high. Kyanite gets its name from the Greek term *kyanos,* which means "dark blue."

STAUROLITE

Staurolite can often be found with kyanite or sillimanite in schists (p. 107) and gneisses (p. 106) that formed during high-pressure regional metamorphism. Its name, based on the unusual cross-shaped crystals that it is famous for, comes from the Greek term *stauros,* which means "cross." While staurolite can form individual crystals shaped like six-sided prisms, it often forms twins with two crystals growing at right angles to each other that form a cross. Staurolite crystals can have a rough texture, but they can also shine like glass. Because it is rich in the elements iron and magnesium, staurolite is most often a dark color such as yellow-brown, reddish brown, or black.

CHONDRODITE

The mineral chondrodite is commonly found in marble (pp. 100–101) formed from dolomitic limestone (pp. 120–121) that has gone through low-temperature and low-pressure metamorphism. It is often found with the minerals diopside (p. 224), spinel (pp. 170–171), graphite (p. 149), and phlogopite (p. 216). Sometimes it forms short stubby prism-shaped or tabular crystals that have a glassy luster. These crystals are commonly yellow, brown, reddish brown, or red. Most often, though, it is found as small crystal grains. Its name comes from the Greek term *chondros*, which means "grain." Chondrodite is the most commonly found member of the humite silicate group of minerals, all of which contain different amounts of the element magnesium in their chemical compositions.

ANDALUSITE

Andalusite gets its name from Andalusia, a region in Spain where it is commonly found. While it occurs in some granite pegmatite rocks, it is much more commonly found in gneisses (p. 106) and schists (p. 107) that have undergone either regional or contact metamorphism. Most often it forms square-shaped prisms that are usually reddish brown or olive green. But it can also be gray or white. The crystals have a glassy luster and are quite hard, but they do have good cleavage in two directions so they often break at square angles. Andalusite crystals are some of the first to form when rocks like shale (p. 127) that are rich in aluminum and silicon are heated at low pressure and temperature.

SILLIMANITE

Sillimanite has the exact same chemical composition as andalusite, but it has a different crystal structure and set of physical properties. It is not quite as hard as andalusite and has only one direction of cleavage. Instead of being shaped like square prisms, its crystals tend to be long, thin needles, and it often produces bunches of thin fibers. Like andalusite, sillimanite comes in shades of brown and green and can also be white. It is found in gneiss (p. 106) and schist (p. 107) that formed at very high temperatures and pressures in sedimentary rocks that were rich in aluminum. Because it requires a great deal of pressure to form, sillimanite is almost always found in rocks that have undergone large-scale regional metamorphism.

A GAGGLE OF GARNETS

Garnets have a split personality. Their awesome colors and glassy luster make for some really cool-looking and inexpensive gemstones. Geologists like them because when they are found in rocks, members of this group of nesosilicate minerals can be very helpful in working out the history of the area. While all garnets have the same general crystal shape, structure, and hardness, individual types have different chemical compositions and colors. Here's a rundown of the six main types of garnet.

GROSSULARITE

Grossular garnets get their name from the Latin term *grossularia,* which means "gooseberry." Typically, this mineral is the same green color as a gooseberry. Grossularite can also be yellow, pink, red, orange, green, and white. Grossular garnets contain the elements calcium and aluminum and are typically found in marble (pp. 100–101) that was formed from limestones that contained silt and clay during contact metamorphism.

PYROPE

Pyrope garnets contain the elements magnesium and aluminum and are usually a deep red to black in color. They can sometimes be purple or rose-colored and even colorless, but these are quite rare. Transparent crystals of pyrope are often used as gemstones. They can often be found as 12-sided crystals called dodecahedrons that look a little like soccer balls in mafic intrusive igneous rocks and schists (p. 107) that have formed under high pressure.

SPESSARTITE

Spessartite garnets contain the elements manganese and aluminum and tend to be mostly orange, brown, and red in color. Even though they are not commonly used as gems, spessartite garnets can produce some really big 12-sided crystals called dodecahedrons that are shaped a little like soccer balls. These crystals are found in granite pegmatites and schists (p. 107) that have formed by regional metamorphism.

UVAROVITE

Uvarovite, which contains the elements chromium and calcium, is the rarest type of garnet. It is almost always an emerald green color, but its crystals are usually too small to allow uvarovite to be used as a gemstone. Uvarovite is often found with the mineral olivine (pp. 234–235) in ultramafic igneous rocks such as peridotite or in metamorphic rocks that have had hot, mineral-rich fluids flowing through them.

ANDRADITE

This common garnet contains the elements calcium and iron and normally comes in shades of yellow, green, brown, and black. When it is transparent and green, it is used as a gemstone called demantoid, but that's pretty rare. Andradite can form in both basalt (p. 90) and granite pegmatites as well as in contact metamorphic rocks that started as limestones.

ALMANDINE

Almandine, the most common type of garnet, contains iron and aluminum. It is usually a deep red to brownish black in color and forms large 12-sided crystals called dodecahedrons that look a little like soccer balls in mica schist (p. 107) and gneiss (p. 106). It may also be found in the igneous rock diorite (pp. 80–81). Clear transparent crystals are often used as gemstones.

DEMANTOID BRACELET

FROZEN IN STONE

Even though many books on rocks and minerals (including this one) list amber in with the minerals, it technically is not a mineral. It does not meet the necessary properties. Minerals have to be inorganic, meaning not made by a living thing. But amber is made by trees, which certainly count as living things. Minerals must also have an internal crystal structure. Even though amber is hard as rock, like glass, it does not form crystals, and the atoms and molecules that make it up are randomly scattered, so it fails this mineral test, too.

So, what exactly is amber, and why is it a gemstone? Amber is fossilized tree resin that has hardened and changed over time. Many sources incorrectly describe amber as being fossilized tree sap. Sap is much different from resin. Sap is thin and is made of water and sugar, and plants use it to get energy. Resin, which is made from other organic compounds including acids and waxes, is much thicker. It is used by trees to seal wounds. Not only will resin not dissolve in water, but it turns hard when it is exposed to air.

Amber is a popular gemstone because it's hard, easy to shape and polish, and looks really cool, especially when it has an insect or baby lizard stuck in the middle of it! While golden amber is the most common color, it can also be brown, red, black, and even blue. And while amber can be completely transparent, it is usually cloudy due to the tiny air bubbles that became trapped inside the resin before it fully hardened. This cloudiness

MUCH OF THE **AMBER** FOUND IN THE WORLD TODAY COMES FROM THE SHORES OF THE **BALTIC SEA** IN **EUROPE** AND IS BETWEEN **40 AND 60 MILLION YEARS OLD.**

Bet You Didn't Know!

The word "electron," which is used to describe the negatively charged particles in atoms and is what makes "electricity" happen, comes from the Greek word *ēlektron*, which means "amber." What's the connection? When you rub amber with wool, it gets a static electric charge, and things like dry grass and feathers will stick to it. It turns out that when amber is rubbed with wool it picks up some extra electrons from the atoms in the wool, which is what gives it the static charge.

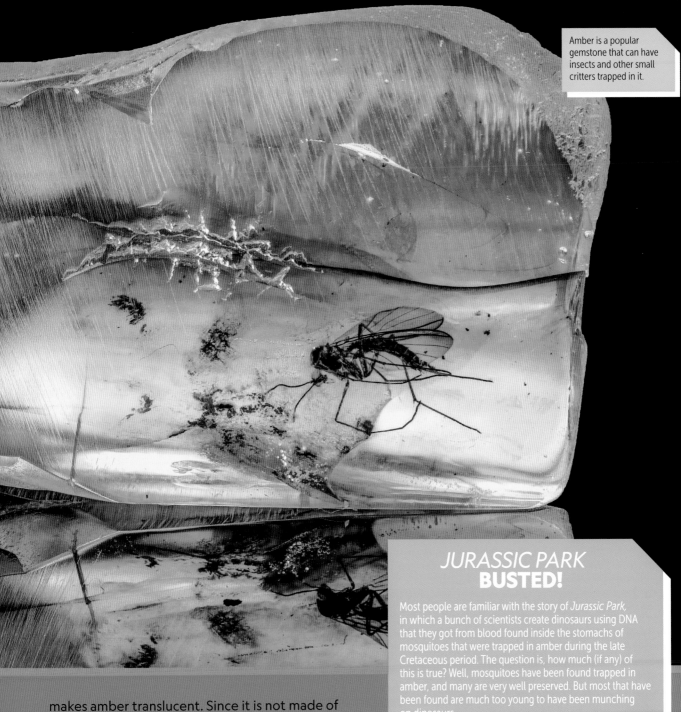

Amber is a popular gemstone that can have insects and other small critters trapped in it.

JURASSIC PARK
BUSTED!

Most people are familiar with the story of *Jurassic Park*, in which a bunch of scientists create dinosaurs using DNA that they got from blood found inside the stomachs of mosquitoes that were trapped in amber during the late Cretaceous period. The question is, how much (if any) of this is true? Well, mosquitoes have been found trapped in amber, and many are very well preserved. But most that have been found are much too young to have been munching on dinosaurs.

Even if we could find a mosquito that lived during the time of the dinosaurs, the chances of finding one that got stuck in resin after it had just chowed down on a *T. rex* would be really low.

The biggest problem, though, would be with the DNA itself. Even if the mosquito were to become perfectly preserved in amber, the DNA inside the blood would begin to break down. Scientists who have run tests on bones and other fossils containing DNA have discovered that after a few million years, most of the DNA chains have completely fallen apart. So, the bottom line is that you don't have to worry about any dinos invading your neighborhood anytime soon!

makes amber translucent. Since it is not made of crystals, amber can't be cut the same way that other gemstones can. Most often it is shaped into beads or pendants or made into bracelets.

Amber does not dissolve in water, but other household chemicals, especially those that include alcohol, can ruin it. Also, it is fairly soft and scratches easily.

WONDERS FROM DOWN UNDER

Some of the most popular and expensive gems aren't dug up in the dirt—they're found under the sea, inside the shells of living creatures called mollusks. So, what are these watery wonders called? Pearls, of course!

Because pearls are made by animals, they cannot be classified as a mineral. However, they are made from calcium carbonate, which is the same substance that the mineral calcite (pp. 190–191) is composed of. Oysters and other mollusks make pearls pretty much the same way they make their shells. Below the shell is an organ called the mantle that produces a substance called nacre. Nacre is made from calcium carbonate in the form of the mineral aragonite and another compound called conchiolin. Together, these two substances produce the shimmering material called mother-of-pearl, which can often be seen lining the inside of a shell.

When some sort of irritant such as a fragment of a shell or a little parasite gets stuck in the mollusk's mantle, the mollusk goes on the defensive and begins covering the irritant with layers of nacre. Over time, these layers build up, creating a pearl that's just waiting to get plucked from the shell!

Not all pearls are perfectly round. Some, called baroque pearls, have an irregular shape. And pearls that are flat on one side because they stay attached to the animal's shell are called blister pearls. While many pearls are white, pearls can also be cream-colored, gray, yellow, green, blue, and even black. The most valuable pearls are those that have the best shine or luster. Value also comes from the "play of colors" that happens when light bounces and bends off a pearl's surface.

THE LARGEST NATURAL PEARL DISCOVERED TO DATE CAME FROM A **GIANT CLAM.** IT WEIGHED IN AT **75 POUNDS (34 KG)** OR ROUGHLY **170,000 CARATS!** THAT'S PROBABLY A LITTLE **TOO BIG FOR A NECKLACE!**

Many people think oysters are the only animals that can produce pearls. However, pearls can be cranked out by many types of mollusks, including clams and mussels. Most pearl-producing mollusks are bivalves, which means they have two shells. But pearls can also be made by certain snails, like the conch.

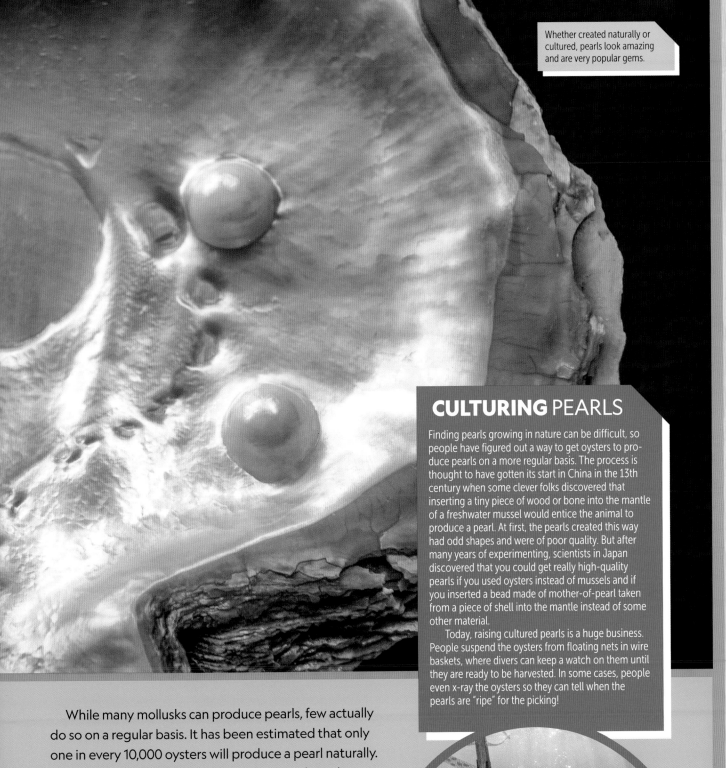

Whether created naturally or cultured, pearls look amazing and are very popular gems.

CULTURING PEARLS

Finding pearls growing in nature can be difficult, so people have figured out a way to get oysters to produce pearls on a more regular basis. The process is thought to have gotten its start in China in the 13th century when some clever folks discovered that inserting a tiny piece of wood or bone into the mantle of a freshwater mussel would entice the animal to produce a pearl. At first, the pearls created this way had odd shapes and were of poor quality. But after many years of experimenting, scientists in Japan discovered that you could get really high-quality pearls if you used oysters instead of mussels and if you inserted a bead made of mother-of-pearl taken from a piece of shell into the mantle instead of some other material.

Today, raising cultured pearls is a huge business. People suspend the oysters from floating nets in wire baskets, where divers can keep a watch on them until they are ready to be harvested. In some cases, people even x-ray the oysters so they can tell when the pearls are "ripe" for the picking!

While many mollusks can produce pearls, few actually do so on a regular basis. It has been estimated that only one in every 10,000 oysters will produce a pearl naturally. This, coupled with the fact that many natural pearl-producing oyster beds have been damaged by fishing and pollution, helps to explain why pearls are so rare. So, if you do happen to discover one of these precious gems, treat it with respect.

CHAPTER EIGHT

ROCKIN'
RESOURCES

Each year, all over the world, mines like this marble quarry in Turkey remove millions of tons of rock to get at the important natural resources that help keep our world running.

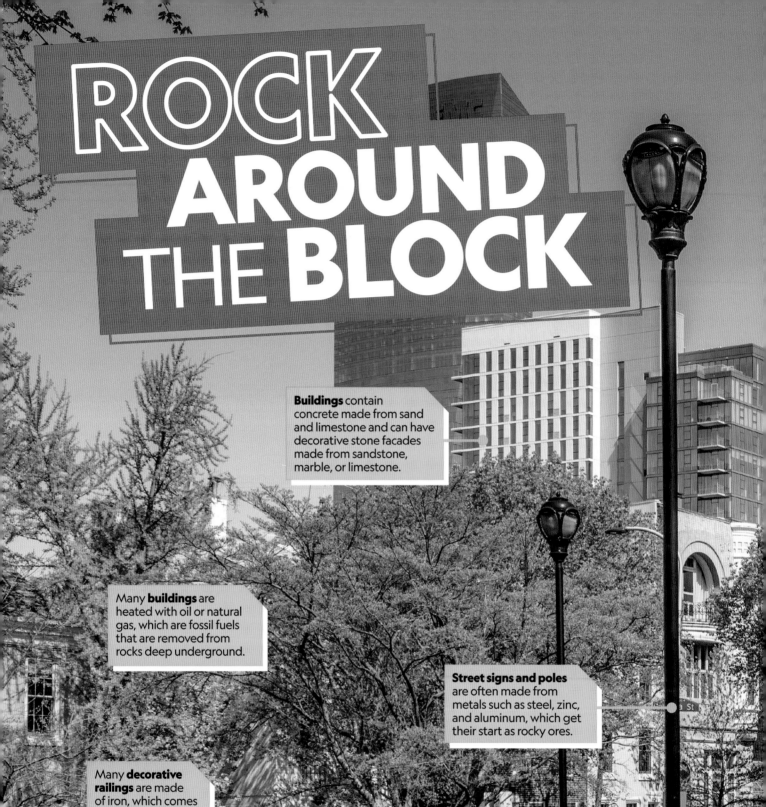

ROCK
AROUND
THE BLOCK

Buildings contain concrete made from sand and limestone and can have decorative stone facades made from sandstone, marble, or limestone.

Many **buildings** are heated with oil or natural gas, which are fossil fuels that are removed from rocks deep underground.

Street signs and poles are often made from metals such as steel, zinc, and aluminum, which get their start as rocky ores.

Many **decorative railings** are made of iron, which comes from different types of rocky ores.

Plastic, which is used in many items, such as baby strollers, comes from oil that formed in rocks underground.

Rocks provide the resources for many of the items we use in our daily lives. Here are a few places that rocks have made an impact on your neighborhood.

Bricks are made from clay that has been baked at a high temperature.

Window glass is often made from quartz sand.

Paving stones can be made from a wide variety of rocks including flagstone, slate, granite, or even marble.

251

ROCKIN' RESOURCES FROM AROUND THE WORLD

People use a lot of different resources to get through their daily lives: industrial metals like steel, aluminum, and copper; fuels like coal and uranium; and even the salt that we use to season our food. All come from rocks and minerals that are found in countries all over the world. Here's a short list of some of the most important rocky resources and the places they are most often found.

COAL

One of the most widely used fossil fuels, coal is used to power factories and generate electricity. Some of the top coal-producing countries are the United States, China, Australia, India, and Indonesia.

SILVER

Silver is used to make decorative items, tableware, jewelry, and even money. Some of the top silver-producing countries are Mexico, China, Peru, Poland, and Chile.

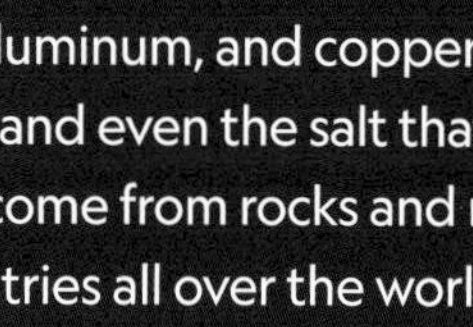

COPPER

Copper is used in a variety of products, from pots and pans to water pipes. It is also the most common metal used for making wiring for electric devices. Top copper-producing countries include Chile, China, Peru, the United States, and the Democratic Republic of the Congo.

Canada

NORTH AMERICA

United States

Mexico

PACIFIC OCEAN

Peru

Brazil

SOUTH AMERICA

Chile

URANIUM

Uranium is used to power nuclear reactors that generate electricity. Top uranium-producing countries include Kazakhstan, Canada, Australia, Niger, and Namibia.

GOLD

Gold is used to make jewelry and decorative items, and it is also found in many electronic devices. Top gold-producing countries include China, Australia, Russia, the United States, and Canada.

IRON ORE

Iron ore is the main source of iron, which is used to make steel for bridges, buildings, cars, and trucks. Top iron ore–producing countries include Brazil, China, Australia, India, and Russia.

SALT

Salt is used for preserving food and melting ice on roads. Salt also tastes great sprinkled on your french fries! Some of the top salt-producing countries are India, China, the United States, Germany, and Canada.

ARCTIC OCEAN

Russia

Poland

Germany →

Kazakhstan

EUROPE

A S I A

China

Niger

Guinea

AFRICA

PACIFIC OCEAN

INDIAN OCEAN

India

ATLANTIC OCEAN

Democratic Republic of the Congo

Indonesia

BAUXITE ALUMINUM ORE

Bauxite aluminum ore is the main source of raw aluminum, which is used to make everything from foil wraps to airplanes. Some of the top bauxite-producing countries are Australia, China, Brazil, Guinea, and India.

Namibia

Australia

MOCK ROCKS

From phenomenal pyramids in Egypt and Mexico to colossal cathedrals like Notre-Dame in France, it's clear that builders of the past loved to use natural stone. But, as cities and towns spread across the globe, natural building stones weren't always available to use. What's a builder to do? Simple: Figure out a way to whip up some rocks of your own!

We don't know for sure who invented the first "brick," but archaeological evidence from the Middle East tells us that people there have been making simple sun-dried mud bricks for about 9,000 years. They likely came up with the idea of making bricks when they realized that the same clay that they molded into pottery could be turned into regularly shaped blocks that could then be stacked, just like cut stone. The big problem with sun-dried brick, however, is that it's not very durable. Even in a dry climate, bricks start to crumble pretty quickly. And, if it rains, nicely shaped bricks will quickly turn back into a pile of mud.

The big breakthrough came around 3500 B.C. when people started using "burnt bricks" in their buildings. We're not sure if they got the idea from their work with pottery or simply got careless one day and dropped some mud bricks into a fire. But they realized that baking bricks in ovens made them much harder. And most important, it allowed them to stand up to the weather.

Today clay is used to make both bricks and their close cousins, tiles, which can be used for almost every type of building project. From huge structures to small patios, walkways, and fireplaces, these artificial "mock rocks" have helped revolutionize the way we build our world.

Bet You Didn't Know!

In addition to inventing the phonograph and perfecting the lightbulb, Thomas Edison came up with his own type of concrete, and he used it to build entire houses. Edison's dream was to provide people inexpensive houses that could be built very quickly. What he would do was build a large mold for parts of the structure into which the concrete was poured, eventually forming the entire house right in place. His idea never caught on, and his cement company eventually went bankrupt.

When natural stone is too expensive or not available, people often turn to brick as a solid alternative for building.

CONCRETE **SOLUTIONS**

While bricks work really well for building walls, they can't do it alone. Unless you have something to hold the bricks together, any sudden jolt will cause your wall to come tumbling down. The stuff that holds bricks together is called mortar, and early on people used mud mortar to hold their mud bricks together. This wasn't all that successful. What was needed was a substance that would go through some type of chemical change so that it could go on wet, but then over time turn solid like rock. The solution was cement.

There are many natural cements that hold sedimentary rocks together, and some of these same chemical compounds have been turned into mortar. These include gypsum, which was used by the ancient Egyptians, and calcium carbonate (or "lime"), which the Romans used. Lime cement was made by crushing limestone and heating it to high temperatures. This created a product called quick lime that could then be mixed with water and fine sand to make mortar or with larger stones to make concrete. Today, most builders use a product called Portland cement, which was invented in the early 1800s. Basically it's a mixture of clay and lime that forms a longer-lasting concrete when it dries.

THE BLACK **PAVING MATERIAL** THAT COVERS MANY **ROADS** IS CALLED **TARMAC.** IT IS SHORT FOR **"TARMACADAM"** AND WAS NAMED FOR **JOHN LOUDON McADAM,** THE SCOTTISH ENGINEER WHO FIRST STARTED **PAVING ROADS** WITH **GRAVEL.**

THAR BE FOSSILS IN THIS FUEL!

Imagine you are living in the distant past, sitting around a campfire that you built using sticks and branches, when suddenly one of the rocks around the edge of the fire starts burning. After you get over your initial shock, you might start looking for more of these strange black rocks that burn. Pretty soon, all of your friends are using them, too, to keep warm, cook food, and maybe even melt some metal and bake some pottery.

OK, maybe it didn't happen exactly that way, but historians think coal was first used to make fire in China well over 3,000 years ago. Over the years, the use of coal has steadily grown, powering steamships and railroads and making the industrial revolution of the 18th and 19th centuries possible. Today, coal is still used as a major energy source, mostly for the generation of electricity, and also in the production of steel and other metals.

Coal, which is classified as an "organic rock," is one of the major exceptions to the rule that rocks are made from minerals. Other than some random grains of sediment that might be trapped in it, coal has no minerals. It's made entirely from partially decomposed plant matter that got buried, squeezed, and heated until most of the water came out of it. Much of the coal being mined and used today came from large treelike ferns, algae, and mosses that grew in great swamps around 300 million years ago during a time called the Carboniferous period. How do we know this? Simple! Geologists have found tons of fossils made

THE COUNTRY OF **ICELAND** GETS ALMOST ALL OF ITS **ELECTRICITY** FROM **FLOWING WATER** AND **STEAM,** HEATED BY **UNDERGROUND MAGMA** THAT **TURNS ELECTRIC GENERATORS.**

Coal is often called a fossil fuel because it is made from the remains of plants. Sometimes, if you are lucky, you can still find the fossils in with the coal!

FUELS FOR THE FUTURE

Fossil fuels such as coal, oil, and natural gas have been the main energy sources people have depended on for more than 200 years. But they are not without problems. Mining coal and drilling for oil cause some serious environmental issues in the areas where they are found. And the burning of these fuels releases carbon dioxide gas and other pollutants into the air. Many scientists believe that it's the buildup of this carbon dioxide in the atmosphere that's causing Earth's surface to become warmer and climates around the planet to change. As a result, many scientists are pushing to limit the use of fossil fuels and, instead, recommend using alternative renewable sources such as wind, water, and solar energy to power our planet.

In many places around the world, the switch is already happening. In parts of Germany and the United States, and in other areas that get a constant breeze, people have built giant wind turbines that make electricity as they spin. Huge dams have been constructed in China, Brazil, and many other countries to create the steady flow of water needed to run hydroelectric power plants. Of these, one of the most promising energy sources comes from photovoltaic solar panels, which can produce electricity wherever the sun shines (that's pretty much everywhere!).

by these plants in the shale rock (p. 127) that forms layers with the coal.

While coal is still technically being produced in some swampy areas today, the process takes hundreds of thousands of years—or, more often, millions of years. And humans are using coal at a much faster rate than the rate it is forming in nature. For this reason, coal is called a non-renewable resource. It's also called a fossil fuel because, you guessed it, it's made from old dead things—some of which have even left us their fossils!

Most of the photovoltaic cells found in solar panels today are made from the element silicon, which is the same stuff that forms the silicon tetrahedrons in quartz (pp. 206–207). So where does all the silicon to make these cells come from? Superpure deposits of quartz sand!

ROCKREATION

Rocks aren't just for reading about. To really appreciate these geologic wonders, you have to experience them firsthand. Here are a bunch of fun activities you can do that allow you to get up close and personal with rocks!

TAKE A HIKE

Whether it's a short trip up a local hillside or a long trip down into the Grand Canyon, one of the best ways to discover the rocky world around you is to get out there and see it for yourself. Make sure you have a trusted grown-up with you when you're out and about!

VISIT A ROCKIN' MUSEUM

Lots of natural history museums have displays of rocks, minerals, and gems. Explore one near where you live or in a city that you are visiting.

CLIMB SOME ROCKS

Climbing walls is nice, but there's nothing like getting up close and personal with the real thing! When climbing, remember safety comes first. Make sure you have a trusted adult with you and all the proper safety gear you need.

MAKE A ROCK, MINERAL, OR FOSSIL COLLECTION

Start your own collection of geologic wonders and let this book be your guide!

BUILD A SANDCASTLE

Spend a day at the beach and put your construction skills to good use building a sandcastle! Just remember to wear your sunscreen!

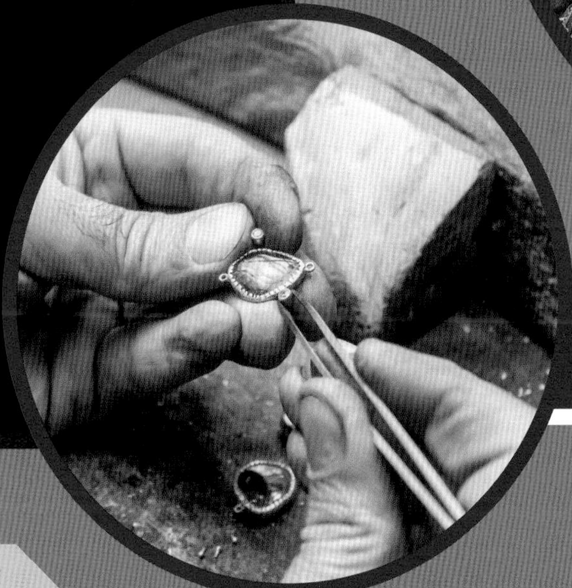

DO SOME LAPIDARY ART

Try making some of your own rock jewelry by taking a class in lapidary.

MAKE SOME POTTERY

Get hands-on with some soft sediments and make some pottery of your own!

DIG FOR DINOS

You don't have to be a paleontologist to join an expedition and help dig for dinosaurs!

MEET A **GEOLOGIST:** A **TALK** WITH DR. ELIZABETH COTTRELL

Dr. Elizabeth Cottrell is a research geologist based in Washington, D.C., where she works for the Smithsonian Institution's National Museum of Natural History. She studies Earth's mantle and what it can tell us about how Earth has changed over time. In addition to conducting both field and lab work, Cottrell also serves as the curator-in-charge of the National Rock and Ore Collections. With more than 325,000 individual specimens, these collections are used by people from all over the world when they are conducting their own geologic research.

What do you find to be the most exciting thing about being a geologist?

I love figuring out how Earth came to be. Earth is unique! It's the only planet (we know of) with continents and oceans. Why don't other planets have those things? I get most excited when I have an idea about how Earth works, and I test my idea by doing experiments in my lab or analyzing rocks. Sometimes my ideas are right! (And sometimes not.) That said, it sure is exciting to take off in a helicopter from the back deck of a boat and fly over the Arctic Ocean to the rim of an active volcano! I got to do this in 2015 when I spent weeks collecting rocks in the Aleutian

Islands, off the coast of Alaska. That trip was almost TOO exciting!

"TOO exciting"? Did anything go wrong?

Overwater helicopter work is really dangerous. Before going, my team had to train for the possibility of a helicopter crash into the ocean. We were strapped into the interiors of scale-model helicopters and planes and plunged into a pool! Without being able to see or breathe, we had to release ourselves from our harnesses, find the door, and get out and to the surface. You couldn't tell which way was up or down! The training really pushed me to my mental and physical limits, and also made me realize how serious the risks were that lay ahead, and that I needed to be a really strong leader for my team. Luckily, we didn't have any major accidents out in the field.

Whew ... that is way too exciting for me! How does your work in the lab support your work in the field?
My field work is motivated by my lab work, and vice versa. Observations in the field result in new hypotheses that we test in the lab. Lab results tell us what we should look for in the field.

So, what types of experiments do you do in the lab?
In my lab I re-create the conditions under volcanoes. Sometimes the experiments are at really high pressures where the solid Earth first melts, and sometimes the experiments are under lower pressure where magmas spend time before erupting. All of the experiments are at a temperature greater than 1400°F (800°C). We can melt rocks this way and then cool them really quickly. We can then study these experiments and compare them to rocks in the field.

What would you consider to be the most important discovery that you've made?
My most "important" discovery might sound pretty boring. I discovered that one of the analytical techniques we use can damage the specimens, and that damage can then result in the collection of bad data. "Bad data" is data that doesn't help you answer the question you were asking—but it might answer a question you didn't even know you had! This work will really help the international community of scientists collect better data.

How important is teamwork when it comes to doing scientific research?
All my work is collaborative; it is all teamwork. My teammates are constantly teaching me new things. They are the real experts—and I am lucky to learn from them. Teammates tell you when you've gone wrong; they check your work; they challenge your ideas.

Where did you grow up, and what inspired you to be a geologist?
I grew up in Essex Junction, Vermont—a town in the far northern part of the state. I grew up spending a lot of time outdoors, hiking, sailing, and cross-country skiing. I had great science teachers in fourth, sixth, 10th, and 12th grades. Science was always my favorite class. I also loved to do science experiments with my dad. I think I decided to be a scientist then.

I started out as a chemistry major in college [at Brown University], but then I took a class about planets in the geology department. That class inspired me to do research in the geology department. A professor in the geology department, named Mac Rutherford, took me into his research group. Professor Rutherford believed in me, and he set me on my path. I did a project on the eruption of Santorini, a volcano in Greece. I conducted experiments to re-create the conditions in the magma chamber beneath Santorini before it erupted catastrophically—creating the iconic horseshoe-shaped caldera we see today.

Do you have any suggestions for kids who might want to one day become geologists themselves?
Get out there and play! Start a rock collection in your backyard. On your next road trip, ask your parents if you can stop at a quarry, mine, or park. Take science classes—and do your best. Geologists can specialize in chemistry, physics, and even biology! So whatever science you love most, you can contribute to our understanding of how Earth works.

GLOSSARY

APHANITIC: a way to describe igneous rocks with small crystals that are not visible to the naked eye

ASTEROID: a large rocky or metallic object that may orbit the sun and can impact Earth, the moon, or other planets

ASTHENOSPHERE: the partially melted layer of the mantle directly below the lithosphere

ATOM: the smallest particle that has all the properties of a chemical element

CAP ROCK: a layer of hard, resistant rock that lies on top of softer rock, often trapping underground deposits of natural gas or oil

CARAT: a unit of weight used to measure gems; equal to 200 milligrams

CHEMICAL BONDING: the process where two or more atoms join together, usually forming compounds

CINDER CONE: a volcano made from layers of volcanic ash and other solid particles that have erupted from inside Earth

CLASTIC: a particle of a rock that has been broken down by weathering, or sedimentary rock made from pieces of rocks that have been joined together again

CLEAVAGE: how a mineral or rock splits or breaks

COMET: an object in space that is made up of ice, dust, and rock fragments joined together

COMPOUND: a substance made from two or more chemical elements bonded or joined together

CONVERGENT BOUNDARY: a place where two tectonic plates collide or come together

CORE: the innermost part of Earth that is made mostly of iron and is divided into a liquid outer core and a solid inner core

CRUST: the outermost layer of Earth directly above the mantle; composed of solid rock

CRYSTAL: a solid substance that has an orderly internal arrangement of atoms that produces a regular shape

CRYSTAL HABIT: the normal shape that a mineral's crystal forms

DENSITY: the ratio between the mass and volume of a substance (density = mass / volume)

DIVERGENT BOUNDARY: a place on Earth's crust where two tectonic plates are pushed apart and, in the process, new crust is created

DUCTILE: a property of a substance that allows it to be drawn out into thin wires

EARTHQUAKE: the shaking of Earth's surface caused by a sudden release of energy

EJECTA: material that is blasted out of a volcano or from a crater after a meteoroid strikes the surface

ELEMENT: a substance with unique chemical properties and which cannot be broken down into a simpler substance by ordinary chemical processes

EPICENTER: the point on Earth's surface directly above the focus of an earthquake

EROSION: the process where rocks and sediment are transported by natural agents such as wind, flowing water, or glaciers

EVAPORITE: a chemical sedimentary rock or mineral that forms as the result of the evaporation of mineral-rich water

EXTRUSIVE IGNEOUS ROCKS: igneous rocks such as basalt that form on Earth's surface

FAULT: a crack in Earth's crust that forms when rocks on either side of the break move past each other.

FELSIC: light-colored igneous rocks that are found mostly on continents and are rich in the minerals quartz and feldspar

FLUORESCENCE: the color of a mineral under ultraviolet light, which is often different than its natural color

FOCUS: the point under the ground's surface where the movement along a fault happens, releasing waves of energy

FOLIATION: parallel alignment of minerals found in certain types of metamorphic rocks

FOSSIL: the remains or traces of a living thing from the distant past preserved in a rock

GEMSTONE (GEM): a precious or semi-precious stone that has been cut and polished and is often used in jewelry

HARDNESS: a measure of how difficult a mineral or rock is to scratch

IGNEOUS ROCK: a rock that is formed by the hardening of melted rock in the form of lava or magma

INDEX FOSSILS: fossils of a particular type of living thing that are found over a wide area and are used to identify the age of the rock that they are found in

INORGANIC: material that has not come from a living thing

INTRUSIVE IGNEOUS ROCKS: igneous rocks such as granite that form under Earth's surface

IRON OXIDE: a chemical compound made from iron and oxygen that is often found in soils and sedimentary rocks and often cements sediments together

KARAT: a measurement of the purity of gold; pure gold is 24 karats

LAPIDARY: the process of cutting and polishing gems, or the person who does the work

LAVA: melted rock that flows over Earth's surface

LAVA DOME: a domelike structure found inside certain volcanoes caused by thick lava pushing up toward the surface

LIMESTONE: a sedimentary rock made mostly of the mineral calcite

LITHIFICATION: the process by which loose sediment turns into a solid sedimentary rock

LITHOSPHERE: the outer layer of Earth, which is made up of the crust and the upper mantle and is broken into large sections called tectonic plates

LUSTER: the way a mineral or rock shines or reflects light

MAFIC: dark-colored igneous rocks that have very little quartz but are rich in olivine and pyroxene minerals and are often found making up the crust under the oceans

MAGMA: melted rock material under the surface of Earth

MALLEABLE: the ability of a substance to be hammered into thin sheets without breaking

MANTLE: the layer of Earth between the crust and the outer core; made of dense rock that in places flows under pressure

MASSIVE: a dense rock that has no layering, or a mineral that forms without distinct crystals

METAMORPHIC ROCK: rocks formed when a preexisting rock changes form due to heat and pressure

METEOROID/METEORITE/METEOR: a meteoroid is a rock from space that forms a streak of light called a meteor when it moves through Earth's atmosphere. If the rock reaches the surface, it is called a meteorite.

MID-OCEAN RIDGE: a mountain chain running along the seafloor formed by magma flowing up through the crust where two tectonic plates diverge or push apart

MINERAL: a naturally occurring inorganic solid with a fixed chemical composition and orderly arrangement of atoms

MOHS SCALE: a scale ranging from 1 to 10 using specific minerals to rank the hardness of all other minerals, with 10 (diamond) being the hardest and 1 (talc) being the softest

MOLECULE: the smallest particle of a chemical compound made of two or more atoms

NEBULA: a large cloud of dust and gas in space

ORE: a mineral or rock that contains useful metals that is mined and processed

ORGANIC: substances that are made by living things

PETROGLYPH: a carving made in a rock

PHANERITIC: a texture found in igneous rocks that has crystals that are large enough to be seen with the naked eye

PLACER DEPOSIT: a deposit of dense minerals that has been concentrated by flowing water or blowing wind

PLATE TECTONICS: the theory that explains how motions of large sections of the Earth's crust cause earthquakes and shape the surface of the planet

PLUTONIC ROCK: igneous rock formed by the cooling of magma below Earth's surface

PRECIPITATE: the process by which minerals that have been dissolved in water form solid particles

RADIOACTIVITY: the energy produced by the natural breakdown of the atoms of certain elements into smaller atoms of a different element

RADIOMETRIC AGE DATING: the process of telling how old a rock or mineral is by using the presence of different amounts of certain radioactive elements; also called absolute age dating

ROCK: a solid substance made of minerals

ROCK CYCLE: the process by which rocks form and then break down to form new rocks of a different type

SEDIMENT: broken pieces of rock or shells that collect on Earth's surface

SEDIMENTARY ROCK: a rock formed at Earth's surface when minerals precipitate from water or when pieces of other rocks, shells, or plant materials get deposited and solidify to form a new rock

SEISMOGRAPH: a device that measures the size and strength of waves moving through the ground as the result of an earthquake

SHIELD VOLCANO: a wide, gently sloping volcano built up by layers of lava that have erupted over time

SILICA: the chemical compound made from the elements silicon and oxygen that forms the mineral quartz

SILICON TETRAHEDRON: a pyramid-shaped compound that forms the basic building block of all the minerals in the silicate class

SONAR: a device that uses sound waves to measure distance between objects and for measuring depth in the ocean

SPECIFIC GRAVITY: the ratio of the weight of a substance compared to the weight of an equal volume of water

SPELEOTHEM: a rock structure formed in caves by the gradual deposition of minerals from water

STALACTITE: a cone-shaped rock structure that hangs down from the roof of a cave

STALAGMITE: a cone-shaped rock structure that points up from the floor of a cave

STRATA: a distinct layer of sediment or sedimentary rock

STRATOVOLCANO: a steep-sided volcano formed by the deposition of alternating layers of lava and volcanic ash; also called a composite cone

STREAK: the colored line left by a mineral after it has been rubbed on a special tile and turned into a powder

SUBDUCTION ZONE: a convergent boundary where one tectonic plate is pushed underneath another tectonic plate and back into the mantle

SUBMARINE TRENCH: a deep, steep valley in the ocean floor near a subduction zone formed when one tectonic plate is forced back into Earth underneath another

TECTONIC PLATE: a section of Earth's litho- sphere that rides on the softer rock of the asthenosphere and slowly moves across the surface of the planet; also called a lithospheric plate

UNIFORMITARIANISM: the geologic principle that states that the processes we see operating on Earth today are the same as those that have happened in the past. Often stated as: The present is the key to the past.

VENT: an opening in a volcano through which lava, steam, and other materials flow out of Earth

VOLCANIC ROCK: igneous rocks that form from the cooling and hardening of lava on the surface of Earth

VOLCANO: a mountain formed by the gradual buildup of lava and other materials from inside Earth

WEATHERING: the process where rocks are worn down and broken on Earth's surface by natural forces such as wind, flowing water, ice, living things, and chemical action

FIND OUT MORE

Great Books, Websites, Videos, and Places to Visit

Books
Absolute Expert Rocks and Minerals by Ruth Strother. National Geographic, 2019

Dirtmeister's Nitty Gritty Planet Earth by Steve Tomecek. National Geographic Kids, 2015

Everything Rocks and Minerals by Steve Tomecek. National Geographic, 2011

1,000 Facts About Dinosaurs, Fossils, and Prehistoric Life by Patricia Daniels. National Geographic, 2020

Rocks and Minerals: A Gem of a Book! by Dan Green and Simon Basher. Kingfisher, 2009

Rocks, Minerals & Gems by Miranda Smith and Sean Callery. Scholastic, 2016

Ultimate Explorer Field Guide: Rocks & Minerals by Nancy Honovich. National Geographic, 2016

Websites
Earthquake Hazards: *usgs.gov/natural-hazards/earthquake-hazards/earthquakes*
This website is a great place to find information on recent or historic earthquakes and selected significant earthquakes

Geology.com: *geology.com*
This website has information on rocks, minerals, and gems as well as explanations of Earth processes, including earthquakes, volcanoes, and plate tectonics.

James Madison University Mineral Museum: *jmu.edu/mineralmuseum*
Maintained by the JMU Department of Geology and Environmental Science, this site features hundreds of excellent photographs showing different minerals and includes their properties.

Mindat.org: *mindat.org*
Mindat.org is the world's largest open database of minerals, rocks, meteorites, and the localities they come from.

Mineralogy 4 Kids: *mineralogy4kids.org*
Presented by the Mineralogical Society of America, this website has lots of great information on the properties of rocks and minerals and how to identify them.

Minerals.net: *minerals.net*
This website presents a searchable database of rocks, minerals, and gems and includes excellent information on their physical and chemical properties, how they form, and where in the world they can be found.

National Park Service Guide to Rocks and Minerals: *nps.gov/subjects/geology/rocks-and-minerals.htm*
This website offers a great overview of how rocks and minerals form, as well as their uses and properties.

Smithsonian Gems Gallery: *geogallery.si.edu/gems*
This gallery contains photos and descriptions of hundreds of gems and jewels from one of the world's largest collections.

Smithsonian Minerals Gallery: *geogallery.si.edu/minerals*
This gallery contains photos and descriptions of hundreds of mineral specimens from one of the world's largest collections.

Smithsonian Rocks Gallery: *geogallery.si.edu/rocks*
This gallery contains photos and descriptions of hundreds of rock samples from one of the world's largest collections.

Volcano World: *volcano.oregonstate.edu*
This website offers a great overview of volcanoes, where and how they form, and current news about volcanic eruptions.

Videos and Movies
Into the Inferno (2016): A look at one scientist's work to study a volcano in North Korea.

Life's Rocky Start (NOVA, 2018): Learn about the connection between rocks and minerals and everyday living things on Earth.

The Minerals Behind Modern Life (NOVA, 2016): Learn how minerals are essential in making everything from skyscrapers to mobile phones.

Supervolcano: Yellowstone's Fury (2013): This movie examines the history of the supervolcano that supplies the energy behind Yellowstone's famous geysers and explores what might happen if it erupts again.

Treasures of the Earth: Gems (NOVA, 2016): Learn how gemstones are formed in the depths of the Earth.

Places to Visit
UNITED STATES: MUSEUMS

American Museum of Natural History, New York City, New York

Colorado School of Mines Geology Museum, Golden, Colorado

Franklin Mineral Museum, Franklin, New Jersey

Houston Museum of Natural Science, Houston, Texas

James Madison University Mineral Museum, Harrisonburg, Virginia

Lizzadro Museum of Lapidary Art, Elmhurst, Illinois

Mineralogical and Geological Museum at Harvard University, Cambridge, Massachusetts

National Museum of Natural History (Smithsonian), Washington, D.C.

Natural History Museum of Los Angeles County, Los Angeles, California

Rice Museum of Rocks and Minerals, Hillsboro, Oregon

San Diego Mineral and Gem Society Museum, San Diego, California

Sterling Hill Mining Museum, Ogdensburg, New Jersey

Vermont Museum of Mining and Minerals, Grafton, Vermont

Wayne State University Geology Mineral Museum, Detroit, Michigan

Yale Peabody Museum of Natural History, New Haven, Connecticut

UNITED STATES: PARKS AND OTHER SITES

Arches National Park, Utah

Bryce Canyon National Park, Utah

Carlsbad Caverns National Park, New Mexico

Crater Lake National Park, Oregon

Glacier National Park, Montana

Grand Canyon National Park, Arizona

Hawaiian Volcano Observatory, Hawaii

Hot Springs National Park, Arkansas

Howe Caverns, New York

Mammoth Cave National Park, Kentucky

Mount Rushmore National Memorial, South Dakota

Mount Washington Observatory, New Hampshire

Petrified Forest National Park, Arizona

Yellowstone National Park, Wyoming

Yosemite National Park, California

AFRICA

Big Hole and Kimberley Mine Museum, Kimberley, South Africa

Bleloch Geological Museum, Johannesburg, South Africa

Egyptian Geological Museum, Cairo, Egypt

Natural History Museum of Zimbabwe, Bulawayo Province, Zimbabwe

Great Pyramid of Khufu, Giza, Egypt

ASIA

Gansu Geological Museum, Lanzhou Shi, Gansu, China

Gargoti Mineral Museum, Maharashtra, India

Geological Museum of China, Beijing, China

Masutomi Geology Museum, Kyoto, Japan

Wulong Furong Cave, Wulong, China

Yamanashi Gem Museum, Yamanashi Prefecture, Japan

AUSTRALIA

Minerals Heritage Museum, Brisbane, Queensland

South Australian Museum, Adelaide, South Australia

Uluru (Ayers Rock), Northern Territory

CANADA

Dinosaur Provincial Park, Alberta

Mineralogy and Petrology Museum, Edmonton, Alberta

Pacific Museum of Earth, Vancouver, British Columbia

Royal Ontario Museum, Toronto, Ontario

Royal Tyrrell Museum, Hanna, Alberta

EUROPE

Cockburn Geological Museum, Edinburgh, Scotland, U.K.

Giant's Causeway, County Antrim, Northern Ireland, U.K.

Lapworth Museum of Geology, Birmingham, England, U.K.

Mineralogical Collection of the Institute of Geosciences, Jena, Thuringia, Germany

Natural History Museum, London, England, U.K.

Oxford University Museum of Natural History, Oxford, England, U.K.

Royal Cornwall Museum, Cornwall, England, U.K.

Sedgwick Museum of Earth Sciences, University of Cambridge, England, U.K.

Stonehenge, Wiltshire, England, U.K.

Werra Kalibergbau Museum and Mine, Heringen, Hesse, Germany

SOUTH AMERICA

Museo Andres del Castillo, Lima, Peru

Museo Mineralis, Cordoba, Argentina

Museum of Science and Technology of the School of Mines, Ouro Preto, Minas Gerais, Brazil

Palacio de Sal (Salt Hotel), Uyuni, Bolivia

INDEX

PHOTO CREDITS

For Patty, Zorida, and Stephen. Thanks for all
your love and support! You folks truly rock! —S.T.

Since 1888, the National Geographic Society has funded more than 12,000 research, exploration, and preservation projects around the world. The Society receives funds from National Geographic Partners, LLC, funded in part by your purchase. A portion of the proceeds from this book supports this vital work. To learn more, visit natgeo.com/info.

For more information, visit nationalgeographic.com or write to the following address:

National Geographic Partners
1145 17th Street N.W.
Washington, DC 20036-4688 U.S.A.

For librarians and teachers: nationalgeographic.com/books/librarians-and-educators

More for kids from National Geographic: natgeokids.com

National Geographic Kids magazine inspires children to explore their world with fun yet educational articles on animals, science, nature, and more. Using fresh storytelling and amazing photography, *Nat Geo Kids* shows kids ages 6 to 14 the fascinating truth about the world—and why they should care. kids.nationalgeographic.com/subscribe

For rights or permissions inquiries, please contact National Geographic Books Subsidiary Rights: bookrights@natgeo.com

Designed by Julie Wood

The publisher would like to acknowledge the following people for making this book possible: Steve Tomecek, author; Priyanka Lamichhane and Libby Romero, senior editors; Jen Agresta, project editor; Angela Modany, associate editor; Sanjida Rashid, art director; Lori Epstein, photo director; Danny Meldung, photo researcher; Mike McNey, cartographer; Joan Gossett, editorial production manager; Alix Inchausti, production editor; Anne LeongSon and Gus Tello, design production assistants; and Michelle Harris, fact-checker.

The publisher would also like to thank Jim Lucarelli for his expert review of the manuscript.

Library of Congress Cataloging-in-Publication Data
Names: Tomecek, Steve, author. I National
    Geographic Society (U.S.) I National Geographic
    Kids (Firm), publisher.
Title: Rockopedia / Steve Tomecek.
Description: Washington, D.C. : National Geographic
    Kids, [2020]. I Includes index. I Audience: Ages 7-10.
    I Audience: Grades 2-3.
Identifiers: LCCN 2019041955I ISBN 9781426339189
    (hardback) I ISBN 9781426339196 (library binding)
Subjects: LCSH: Rocks--Identification--Juvenile
    literature. I Rocks--Miscellanea--Juvenile literature.
    I Geology--Juvenile literature.
Classification: LCC QE432.2 .T663 2020 I
    DDC 551.02--dc23
LC record available at lccn.loc.gov/2019041955

Printed in Hong Kong
20/PPHK/1